# The
# Soul of Soil

# The Soul of Soil

## A Guide to Ecological Soil Management

Third Edition

Grace Gershuny & Joseph Smillie

agAccess
Davis, California
1995

First Edition 1983
Second Edition 1986
Third Edition 1995

Printed in the United States of America

**agAccess** is an agricultural and horticultural publishing company dedicated to enhancing sustainable food production through the worldwide publication and distribution of high quality, practical information. We publish scientific, technical and popular books, and welcome proposals for new publications.

For more information and a free catalog, please contact us.

**agAccess,** 603 Fourth St., Davis, CA 95616 (916) 756-7177

ISBN 0-932857-16-7
Library of Congress Catalog Card Number: 95-76375

Cover illustration by Stewart Hoyt.
Designed by Timothy Rice.

Printed on recycled paper.

This work is dedicated
to life on earth
and all soil lovers.

❖

# Contents

## List of Tables



## Appendices

# Foreword

*A thin layer of earth, a few inches of rain, and a blanket of air make human life possible on our planet. These essential resources must be available to provide the good life for our children and for future generations.*

John F. Kennedy

*...I am intrigued by the thought that good soils make good people...*

Hans Jenny

When *The Soul of Soil* was first published back in 1983 it was almost an anomaly. It suggested that "the first step toward effective ecological soil management is an appreciation of the complex, living system known as soil," that "to understand soil is to be aware of how everything affects and is affected by it," and that "we are all part of the soil ecosystem." This was far from the conventional wisdom that soil was really just an inert material necessary to hold a plant in place while we artificially applied the required minerals.

True, soil can be viewed from the perspective of its component parts—its physical, biological and chemical properties. But soil is more than that. Taken as a whole, soil is a living organism that is intimately connected to all of life, including the human species.

Gradually soil scientists are coming to the realization that soil really does have "soul," a "principle of life," as the *Oxford English Dictionary* defines soul. Scientists are coming to this conclusion because the reductionist perspective of manipulating soil's isolated elements is getting us into trouble. The reductionist view blinds us to the fact that soil management practices which produce maximum yields can still erode soil at unprecedented rates. It often devalues the role of micronutrients that are essential to human nutrition. It sometimes ignores the role of soil fauna, and the wondrous relationships between soil's living populations and soil tilth and soil health. It has clearly disregarded the connection between soil and ourselves.

Today the suggestion that soil may have soul no longer strikes us as odd. Soil scientists everywhere are now discussing soil quality and soil health. The notion that soil quality must be taken as seriously as water or air quality is now acknowledged by the scientific community. During the era of soil science when dirt was just an anchor for plants, soil was considered "fertile" if it simply contained adequate amounts of three macronutrients—nitrogen, phosphorous and potash. Gradually soil scientists began to recognize the importance of micronutrients, the importance of soil structure and its relation to compaction and erosion, and the importance of earthworms. Still later the importance of the "rate of organic matter turnover" was recognized. Now "soil quality" and "soil health"—the total soil organism and its place in a healthy ecosystem—are preoccupying soil scientists.

Recently ethicists have joined the chorus. They are reminding us that we have an interdependent relationship with the soil and that this raises rich questions going to the core of who we are as human beings sustained by a planet that nurtures our lives. This suggests that when we contemplate the question of what kind of agriculture we want for the future, we have to ask more than the "scientific" questions surrounding food safety and environmental protection. We have to ask what our relationship is to the soil, and how that relationship, or lack of relationship, affects who we are as a people.

This issue was framed clearly by Ivan Illich and some of his colleagues. They said:

> The ecological discourse about planet earth, global hunger, threats to life, urges us to look down at the soil, humbly, as philosophers. We stand on soil, not on earth. From soil we come, and to soil we bequeath our excrements and remains. And yet soil—its cultivation and our bondage to it—is

> remarkably absent from those things clarified by philosophers in our western tradition.
>
> As philosophers, we search below our feet because our generation has lost its grounding in both soil and virtue. By virtue, we mean that shape, order and direction of action informed by tradition, bounded by place, and qualified by choices made within the habitual reach of the actor; we mean practice mutually recognized as being good within a shared local culture which enhances the memories of a place.
>
> We note that such virtue is traditionally found in labor, craft, dwelling and suffering supported, not by an abstract earth, environment or energy system, but by the particular soil these very actions have enriched with their traces. And yet, in spite of this ultimate bond between soil and being, soil and the good, philosophy has not brought forth the concepts which would allow us to relate virtue to common soil, something vastly different from managing behavior on a shared planet.
>
> We were torn from the bonds to soil—the connections which limited action, making practical virtue possible—when modernization insulated us from plain dirt, from toil, flesh, soil and grave.
>
> To speak of friendship, religion, and joint suffering as a style of conviviality—after the soil has been poisoned and cemented over—appears like academic dreaming to people randomly scattered in vehicles, offices, prisons and hotels.[1]

This suggests that our modern industrialized society has brought about a divorce; we have become divorced from the soil. We will now pay a great price for that divorce—an alimony paid in the currency of the environment that will extend through countless human generations.

---

[1] Sigmar Groeneveld, Lee Hoinacki, Ivan Illich, et. al., "The Earthy Virtue of Place," *New Perspectives Quarterly,* (Vol. 8, # 1, Winter, 1991), p. 59.

Until we heal that divorce and become lovers of the soil again, relate soul to soul, many of our social problems will go unsolved —including our food safety and environmental protection problems. We cannot be sustained as a people and a culture divorced from the soil.

This little volume is an invitation to all of us to *become lovers of the soil again* as a way of healing our soils and our souls. This requires that we go beyond the loveless marriage with dirt and establish a relationship of true mutuality and love, that we husband it in the deepest sense. *The Soul of Soil* helps us to understand the complexity of the soil and how to care for it. It is a manual of husbandry. It teaches us how to tend the soil properly.

We need this intimate relationship with the soil because the soil is the connection to ourselves. From soil we come, and to soil we return. If we are disconnected from it we are disconnected from ourselves. We are aliens adrift in a place without fulfillment, without being well. It is the soil that helps us to understand the answer to the age-old question about the meaning of life—the self-limitations of life, its cycles of death and rebirth, the interdependence of all species, the "communion of subjects" (as Thomas Berry puts it[2]) of all of the earth community. To be at home with the soil is truly the only way to be at home with ourselves, and therefore the only way we can be at peace with the environment and all of earth's species. It is, literally, the common ground on which we all stand.

Why have we become divorced from it? If soil is the source of life, why have we become alienated from it? Why do we treat it like dirt? The answer lies in our arrogance. Arrogance prevents us from acknowledging our tie to dirt. We are such an elegant, smart, inventive species—it is simply beneath us (no pun intended) to entertain the notion that we are somehow inextricably wed-

---

[2]Brian Swimme and Thomas Berry, *The Universe Story,* (New York: HarperCollins, 1992).

ded to dirt. But the fact remains that we are tied to it. My friend, John Pitney who has made a career of writing songs about the care of the land, has put it eloquently: "The fact that we are not *now*...dirt, is only temporary."

There are limits. Many of us would like to insulate ourselves from those limits. Becoming lovers of the soil puts us back in touch with those limits. The fact is, we don't like being tied to the soil's limits because they remind us of our own limits. Most of us like the idea that modernization insulates us from dirt, toil, flesh, and grave.

*The Soul of Soil* can help us to rebuild our connections to the soil by helping us to understand it as a living system and to care for it as a part of ourselves.

In addition to the soil husbandry practices suggested in *The Soul of Soil* we might develop rituals of consent that gardeners and farmers could perform before they prepare the soil for seeding. Many ancient cultures, including Native American, practiced such rituals, perhaps with good reason. A regular ceremony during which growers ask the soil's consent for what they are about to do in preparing it for seeding could serve to remind them that they are holding life in their hands. If I ask the soil's consent for what I am about to do, I am more likely to attend to its needs, to be in touch with its cycles, and to invoke a cultural memory of caring for the soil in that place. It might be one way of beginning the long journey back to loving the soil again.

Frederick Kirschenmann

# Acknowledgments

This book was first published in 1983 by the Extension Service, University of Vermont, having been undertaken as part of a Masters program in Extension Education by Grace Gershuny. Joe Smillie, originally a reviewer and mentor of this work, significantly expanded and improved the original in 1986, resulting in a second edition which received widespread attention. This fruitful collaboration has continued, despite increasingly busy work schedules and life changes, and the third edition of *The Soul of Soil* is now a reality. This new version reflects updated research data about soil management, as well as a broadened geographical scope to increase its applicability nationally and internationally.

Many people have helped us at different stages of drafting, self-publishing and revision. Among those who have helped with manuscript review and suggestions are Bart Hall-Beyer, Robert Parnes, Will Brinton, Miranda Smith, Zea Sonnabend, Eric & Beth Ardapple-Kindberg, and Win Way. Bill Wolf, of Necessary Trading Company, has been a reliable source of advice, as have Stuart Hill and associates at the Ecological Agriculture Project, Canada. Early drafts were reviewed by Bill Kruesi, Fred Magdoff, Bill Murphy, Bill Liebhardt, and Miriam Klein. Fred Franklin challenged us to stay on the cutting edge.

Karen Van Epen of agAccess has been most patient in waiting for bits and pieces of copy and having to track us down at different ends of the globe. Some of the updated information included here was unearthed in the process of writing a new book for Rodale Press, Inc., entitled *Start With the Soil* (Grace Gershuny, 1993). Thanks are due to her editor, Sally Roth, and the staff at Rodale.

Our best advisors have been the many organic farmers we have been privileged to visit, work, play, and live with. Friends and colleagues of NOFA (Northeast Organic Farming Association), as well as the many similar groups around the nation and the world, have offered continual inspiration.

Personal, moral and creative support came from our families. Stewart Hoyt, Grace's ex-husband, did the cover and illustrations for the second edition, most of which are included here. Susan Boyer, Joe's wife, helped produce the second edition. Friends in Jamaica, including current spouse Bentley Morgan, helped Grace gain new perspectives and learn a little about tropical agriculture. Our children, Opal Hoyt and Megan, Adrienne and Ellen Smillie are who it's all for.

# About the Authors

Grace Gershuny has been active in alternative agriculture for over twenty years, as an organic farmer, organizer, author, and educator. She has taught agriculture at the Institute for Social Ecology and Goddard College, Plainfield, Vermont, and for many years was closely associated with the Northeast Organic Farming Association (NOFA). Ms. Gershuny gained wide recognition as editor of *Organic Farmer: The Digest of Sustainable Agriculture,* published from 1990-1994. She is author of *Start With the Soil* and is coauthor of *The Rodale Book of Composting.*

Ms. Gershuny has worked in every aspect of organic certification program development and management, and has recently joined the staff of the U.S. Department of Agriculture's National Organic Program to implement its accreditation program for organic certification agencies.

Joseph Smillie has been a consultant in ecological agriculture since 1977. He is coauthor of *The Orchard Almanac: A Seasonal Guide to Healthy Fruit Trees; Guidelines for the*

*Organic Foods Industry;* and *Rodale's Chemical-Free Yard & Garden.* He has been instrumental in developing certification systems in North America and is a founding member of the Independent Organic Inspectors Association (I.O.I.) and the International Inspectors Group (I.I.G.).

A founding member of the Organic Trade Association, Mr. Smillie is currently a board member and co-chair of the Quality Assurance Council. As a consultant to organic food companies and a certification inspector, he has worked with farmers and orchardists around the world. He somehow finds time to tend his garden, and homesteads with his family on a plateau in Erle, Québec.

The authors also collaborated on drafting *Guidelines for the Organic Industry* for the Organic Foods Production Association of North America (OFPANA) in 1986.

# 1
# Introduction

There are as many different concepts of farming as there are farmers. The last fifty years have witnessed the rapid industrialization of agriculture, with spectacular improvements in yields per acre and per person hour. But these achievements have come at high social and environmental cost, leading to widespread doubt about the wisdom of treating agriculture as just another industrial process. Today, biotechnologists are attempting to speed up this revolution even more—if the ultimate scenario of industrial agriculture is the farmerless farm, that of biotechnology could well be the farmless food system.

The farm crisis of the eighties has not disappeared. Farms are still getting fewer and larger overall, while chronic problems of "surplus" and low prices plague those who hold on. Global market forces are subjecting small farmers everywhere to the whims of giant commodity traders. Topsoil in the cornbelt is eroding faster than during the dustbowl days of the thirties, and ecological destruction threatens farmland throughout the Third World and former Soviet bloc. Farmworkers worldwide face serious health problems from exposure to pesticides—many banned in the U.S.—while pests grow increasingly resistant.

Early critics of industrialized agriculture, such as Sir Albert Howard and J. I. Rodale, used the term "organic" to describe farming methods based on living processes. That word has stuck, especially in the minds of consumers, who now demand "organically grown" foods when they are looking for products grown without synthetic pesticides in an ecologically sound manner. "Biological" is also a popular term, especially in Europe, and avoids the confusion caused by those who point out that all carbon-based compounds—including pesticides, herbicides, and other petroleum products—are technically defined as "organic." Some like to refer to "regenerative" agriculture to emphasize the need to repair some of the ecological damage done to soil and water.

In recent years, many of these concepts have become accepted within the established agricultural research and education institutions. In fact, "sustainable" agriculture, once a nearly interchangeable term with "organic," has become the new mandate for many federal and state agriculture programs, in the U.S. and worldwide. The 1992 United Nations Earth Summit, in fact, established sustainable agriculture as one of the key areas of emphasis in order to reverse the planet's current course toward ecological destruction. The term "sustainable" eludes definition and stands for a set of general principles rather than a marketplace label. "Low-input," referring to reduced reliance on off-farm inputs, has also been used to characterize alternative agriculture in a way that emphasizes the economic benefit of decreased production costs.

While we have worked with constituencies who identify with all of these different terms, the one we prefer is "ecological" agriculture or "agro-ecology." The science of ecology emphasizes the interrelations between all living things and their environment, which includes social, economic, and political relationships. As this understanding takes root on more working farms, it will come to be known simply as "good farming": that which is based on resource-efficient and ecologically harmonious methods.

Whichever terminology feels most comfortable to you, soil management is the basis of farm productivity, and therefore the cornerstone of any ecologically sound approach to farming. The most crucial element in any plan for conversion is the soil management plan. Many other farm decisions, including crop selection and equipment purchase, must be made with consideration of soil management needs. And once you have grasped the complex dynamics of the soil ecosystem, you will more easily perceive other practical applications for agro-ecological principles. While reliance on purchased inputs is reduced, increased skill, observation and management capabilities are required in order to eliminate the need for the fast-acting remedies relied on in conventional systems.

There is continuing controversy about what methods and materials constitute ecological or sustainable agriculture, but there are now fairly specific regulatory stipulations about practices permitted for organic farming. While we have participated in many of these discussions, this book is not intended as a set of authoritative prescriptions. Rather, we hope it will be used as a reference and an aid to planning, adapted to the unique needs of the reader. More information related to requirements for organic certification is found in Chapter 5, and occasional comments are included throughout the text as to the acceptability, according to legal definitions of organic, of various materials and practices under discussion.

It is often difficult for those wishing to practice ecological methods to find information and advice about how to do it. Traditional sources such as Extension and farm suppliers, although they have made considerable progress in recent years, are still limited by their lack of practical knowledge, their limited budgets, their conflicting desire for increased sales. Nonprofit nongovernmental organizations are generally underfunded and poorly staffed, often lacking basic agricultural knowledge. Much can be learned from other farmers, trial and error, and diligent research, but there is a real need for practical guidelines to help make sound soil management decisions based on ecological principles. That is the purpose of this book.

This book was originally written from a northeastern U.S. bioregional perspective. This edition has been considerably revised to reflect our wider experience in other regions of this country and the world, and to make the information more directly applicable in other places. The principles of ecological soil management can be used by farmers anywhere to improve the health of their soil and the productivity of their farm operations.

## *Four Definitions of Ecological Agriculture*

The fundamental concepts of ecological agriculture have been well summarized in the following statements.

**From the Organic Farmers' Associations Council**
***Principles of Organic Agriculture***

Organic farming practices are based on a common set of principles that aim to encourage stewardship of the earth. Organic producers work in harmony with natural ecosystems to develop stability through diversity, complexity, and the recycling of energy and nutrients. They seek to:

- Provide food of the highest quality, using practices and materials that protect the environment and promote human health.
- Use renewable resources and recycle materials to the greatest extent possible, within systems that are regionally organized.
- Maintain diversity within the farming system and its surroundings, including protection of plant and wildlife habitat.
- Provide livestock and poultry with conditions that meet both health and natural living requirements.
- Seek an adequate return from their labor, while providing a safe working environment and maintaining concern for the long-range social & ecological impact of their work.

❖ ❖ ❖

**Organic Trade Association (OTA)**
***Guidelines for Certification Organizations***

**Organic foods** are produced under a system of ecological soil management which relies on building humus levels through crop rotations, recycling organic wastes, and applying balanced mineral amendments. This, along with the use of resistant varieties, minimizes disease and pest problems. When necessary, mechanical, botanical, or biological controls with minimum impact on health and environment may be used. Organic foods are processed, packaged, transported and stored to retain maximum nutritional value without the use of artificial preservatives, coloring or other additives, irradiation or synthetic pesticides.

**Sustainable agricultural systems** are based on ecological soil management practices which replenish and maintain soil fertility by providing optimum conditions for soil biological activity. They aim to reduce the use of off-farm inputs, environmental and health hazards associated with agricultural chemicals, and reliance on non-renewable resources. Sustainable agricultural systems are modeled on natural ecosystems in which diversity, complexity, and the recycling of energy and nutrients is essential.

**Basic Tenets of Organic Agriculture from the USDA's *Report & Recommendations on Organic Farming***

1. **Nature is capital:** Energy-intensive modes of conventional agriculture place man on a collision course with nature. Present trends and practices signal difficult times ahead. More concern over finite nutrient resources is needed. Organic farming focuses on recycled nutrients.
2. **Soil is the source of life:** Soil quality and balance (that is, soil with proper levels of organic matter, bacterial and biological activity, trace elements, and other nutrients) are essential to the long-term future of agriculture. Human and animal health are directly related to the health of the soil.
3. **Feed the soil, not the plant:** Healthy plants, animals and humans result from balanced, biologically active soil.
4. **Diversify production systems:** Overspecialization (monoculture) is biologically & environmentally unstable.

*Continued on next page.*

5. **Independence:** Organic farming contributes to personal and community independence by reducing dependence on energy-intensive agricultural production and distribution systems.
6. **Anti-materialism:** Finite resources and Nature's limitations must be recognized.

❖ ❖ ❖

**Earth Summit Alternative Nongovernmental**
***Sustainable Agriculture Treaty***

Sustainable agriculture is a model of social and economic organization based on an equitable and participatory vision of development which recognizes the environment and natural resources as the foundation of economic activity. Agriculture is sustained when it is ecologically sound, economically viable, socially just, culturally appropriate and based on a holistic scientific approach . . . Sustainable agriculture respects the ecological principles of diversity and interdependence and uses the insights of modern science to improve rather than displace the traditional wisdom accumulated over centuries by innumerable farmers around the world.

# 2

# Understanding the Soil System

The first step toward effective ecological soil management is an appreciation of the complex, living system known as soil. And to understand soil is to be aware of how everything affects and is affected by it. We are all part of the soil ecosystem.

Soil fertility can be described as its capacity to nurture healthy plants. Sustainable agriculture aims to protect the soil's ability to regenerate nutrients lost when crops are harvested—without dependence on "off-farm" fertilizers. This regenerative capacity, in turn, depends on the diversity, health and vitality of the organisms that live, grow, reproduce, and die in the soil. Through the activities of soil microbes—which can number in the billions in every gram of healthy topsoil—the basic raw materials needed by plants are made available at the right time, and in the right form and amount.

**The basic aim of ecological soil management is to provide hospitable conditions for life within the soil.**

Your farm is both the product and producer of soil. Consider your farm to be a living organism that achieves its greatest long-term productivity when its natural cycles and processes are enhanced. Short-cutting these cycles for short-term control or economic gain will eventually bear out the ecological maxim, "The creature that wins against its environment destroys itself."

The place to start is where you are. Thousands of soil types have been named, classified and described. Knowing their names can tell you a lot about their general characteristics; but, like any living creature, each individual is unique. Find out what soils live in your area, how they are classified and described by soil scientists, and how that compares with what you observe about them yourself.

Soil classification schemes organize soils according to their different qualities, based on the kinds of minerals they contain, how they were formed, and various physical characteristics. The individual character of any soil arises from a combination of factors inherent to its particular geographic region (see Table 1):

**Climate:** Temperature and precipitation affect the rate of organic matter accumulation and the presence of soluble soil minerals. For example, more organic matter accumulates where decomposition is slow due to cooler temperatures, while high rainfall leaches mineral nutrients from topsoil.

**Native vegetation:** Grassland, forests and transition zones each affect soil development in a different way. Leaf litter from pine forests, for example, increases soil acidity. The particles of soils developed under grasslands are usually bound into stable aggregates by the activity of the plentiful microorganisms and roots found there.

**Parent material:** Underlying rock types from which it was formed determine a soil's mineral content and basic textural qualities. Limestone bedrock, for instance, helps counteract soil acidity. Red soils indicate that the parent material and derived soil is rich in iron. Volcanic ash produces soils heavy in amorphous clays.

**Topography:** Soil may be eroded from slopes and deposited in lowlands. The legendary fertility of river valleys such as the Nile resulted from deposits of rich sediment carried from the highlands, while mountain farmers worldwide have problems holding onto precious topsoil.

**Time:** How long the native rock has been subject to weathering influences the availability of minerals and extent of humus development. Young soils, such as those in Hawaii and other areas of volcanic activity, may be low in clay content, which is produced by the chemical effects of weathering on parent rocks.

**Glaciation and geologic activity:** In the north-temperate region, advance and retreat of the glaciers, most recently a mere 12,000 years ago, has had a significant effect on soil formation and quality. Volcanic activity has left nutrient-rich lava deposits in many areas.

Soils worldwide have been classified into ten major orders (Table 2). In humid temperate regions such as the northeastern United States, where forests are the predominant natural vegetation, the soil order of Spodosols is most common. These soils are generally formed from coarse-textured parent material, and tend to be quite acid and low in mineral nutrients. Prairie soils, which have developed under flat, grass-covered areas with modest rainfall, are classified as Mollisols. They are among the most naturally productive soils, with high native organic matter and mineral content. In tropical regions with very high seasonal rainfalls, the heavily leached Ultisol soils also tend to acidity. The Sahara, Gobi and Turkestan Deserts, as well as South and Central Australia and the American Southwest are largely comprised of Aridisols. If irrigated, they can be productive, but great care must be taken to prevent toxic accumulations of soluble salts.

Each order is further broken down into suborders, great groups, and subgroups. Beyond this, soils are described in terms of families, associations and series, which provide more information about their plant growth characteristics, organic and mineral content, structure, drainage and color. Series are often named after the places—towns, rivers, or counties—where they are located.

Your local Extension or Soil Conservation Service office can probably give you a soil map for your land. They can also show you your county's soil survey, which provides detailed information on local soils and their best uses, as well as helpful climatological data.

*Table 1. Environmental influences on soil.*

| Factor | Influence | Example |
|---|---|---|
| GEOLOGICAL | | |
| Parent material | Determines physical and chemical properties of raw geological material that will form soil. | Limestones delay acidification in humid regions; clays are prone to drainage problems; sands are excessively leached and drained. |
| Glaciation | Bedrock was exposed, parent materials redistributed, and debris deposited by migrating glaciers. | Glacial moraines, such as sand and gravel pits; unusually rocky fields. |
| Weathering | Wind and water cause erosion, and affect soil's sedimentation history. | Rich alluvial soils (river bottom land), deposited by spring run-off. |
| GEOGRAPHICAL | | |
| Altitude & Topography | Affect accessibility, drainage, warmth, and protection from the weather. | Lowlands with clay parent materials are poorly drained. |
| Settlement patterns | Affect economic value and pollution of soil. | Industrial acid rain alters soil chemistry. |
| Watershed | Provides drainage basin for rivers and a stable source of water. | Water source independent of rainfall. |

| Factor | Influence | Example |
|---|---|---|
| CLIMATOLOGICAL | | |
| Wind patterns | Determines severity of erosion and desiccation problems of exposed land. | Windbreaks on one edge of a cultivated field can preserve soil. |
| Precipitation | Determines rate and amount of water entering the soil. | Excessive rainfall dissolves and carries nutrients away from soil. |
| Temperature | Affects rate of chemical reactions and biological activity. | Decay organisms decompose raw organic matter more quickly during warmer months. |
| Microclimate | Localized areas that are warmer or cooler, wetter or drier, than surrounding territory will change soil. | Warmer regions near bodies of water protect cold-sensitive soil organisms. |
| BIOLOGICAL | | |
| Wildlife populations | Responsible for nutrient recycling, burrowing and digging in soil. | Declining predator populations can lead to more groundhog burrows. |
| Ecological succession | Determines dominant type of vegetation best suited for soil type and climate. | Abandoned fields support poplar and other fast-growing shrubs; these, in turn, support forest growth by adding nutrients and organic matter as they die and decompose. |
| Human use | Affect soil through cultivation, fertilization, altered drainage patterns, irrigation, non-agricultural development, and waste disposal. | Chinese farmers reclaimed severely eroded hillsides with intensive terracing and waste recycling. |

*Table 2. The ten major soil orders (adapted from Brady).*

| | |
|---|---|
| Entisols: | Recently formed mineral soils with little evidence of horizon formation. Found in a wide range of climate zones, including the Rocky Mountains, the Sahara Desert, Siberia, and Tibet. May be highly productive, but most are relatively barren. |
| Vertisols: | Mineral soils with a high content of swelling-type clays, which in dry seasons cause the soils to develop deep cracks. Found in some areas of the southern U.S., India, Sudan, and eastern Australia. Their physical properties make them difficult to till and cultivate. |
| Inceptisols: | Young soils with limited horizon formation. May be very productive, as those formed from volcanic ash. Found in the Pacific Northwest (U.S.), along the Amazon and Ganges rivers, north Africa, and eastern China. |
| Aridisols: | Mineral soils found mostly in dry climates. Productive only if irrigated, and may become saline. Found in the southwestern U.S., Africa, Australia, and the Middle East. |
| Mollisols: | Characterized by a thick, dark surface horizon, they are among the world's most productive soils, with high natural fertility and tilth. Generally found under prairie vegetation, such as the Great Plains (U.S.), the Ukraine, parts of Mongolia, northern China, and southern Latin America. |
| Spodosols: | Mineral soils characterized by distinct horizons, including subsurface organic matter, and aluminum and sometimes iron oxides. Coarse textured, readily leached, and tending to be acid, they occur mostly in humid, cold temperate climates, generally under forests. Can be very productive if properly fertilized. |
| Alfisols: | Moist mineral soils with high base status and presence of silicate clays. Found mostly in humid regions under deciduous forest or grass, including parts of the U.S. Midwest, northern Europe, southern Africa, and southeast Asia. Highly productive, good nutrient levels and texture. |
| Ultisols: | Moist soils which develop under warm to tropical climates. Highly weathered, acidic, with red or yellow subsurface horizons. Found in the humid southeast U.S., southeast Asia and southern Brazil. Can be highly productive, with good workability. |
| Oxisols: | The most highly weathered soils, with a deep subsurface horizon of iron and aluminum oxides. High in clay, commonly deficient in phosphorus and micronutrients, they are most common in the tropics on old land surfaces. Not well adapted to mechanized farming, they have been little researched. |
| Histosols: | Organic soils which have developed in a water-saturated environment, with at least 20 percent organic content. Can be very productive if drained, especially for vegetable crops. |

## Organic Matter & Humus

Soil health and humus are indivisible—health is the vitality of the soil's living population, and humus is the manifestation of its activities. As the cornerstone of the soil ecosystem, humus influences and is influenced by every other aspect of the soil. Building soil humus improves its physical and chemical properties as well as its biological health.

All humus is organic matter—but not all organic matter is humus. Raw organic matter consists of the waste products or remains of organisms that have not yet decomposed. Humus is one form of organic matter which has undergone some degree of decomposition. There is no hard and fast dividing line, but a continuum, with fresh, undecomposed organic materials—manure, sawdust, corn stubble, kitchen wastes, or insect bodies—at one end, and stable humus, which may resist decomposition for hundreds of years, at the other. Table 3 summarizes the attributes of different types of organic matter and humus.

Humus is dark brown, porous, spongy and somewhat gummy, and has a pleasant earthy fragrance. Chemically, it is a mixture of complex compounds, some of which are plant residues which don't readily decompose, such as waxes and lignins. The rest are gums and starches synthesized by soil organisms, primarily bacteria and fungi, as they consume organic debris. Humus is highly variable in its composition, depending on the nature of the original material and the conditions of its decomposition.

Humus is actually more a generic term than a precise one. Its qualities will reflect different origins and composition. Just as wine can vary widely in quality, so can humus. And, just as different wines are suitable for different culinary purposes, the varieties of humus serve varying soil functions.

Several classification schemes for humus have been suggested. Theories differ as to how it is formed, why it behaves as it does, and how it should be measured. Humus which can still

*Table 3. The nature and function of organic matter and humus.*

| | **Raw organic matter** | **Effective humus** | **Stable humus** |
|---|---|---|---|
| NATURE | | | |
| Source | Wastes, residues, and remains of living organisms. | Decomposed raw organic matter. | Decomposed raw organic matter or effective humus. |
| Composition | Complex organic compounds, such as proteins, cellulose, lignins, fats, starches and sugars. | Characterized by high ratio of fulvic acids (small, soluble molecules). | Mostly long-chained humic acids, or humins bonded to clay particles. |
| Characteristics | Heterogeneous, coarse, lumpy material. | A colloid; more homogeneous in texture and color. | Homogeneous; resistant to chemical action. |
| FUNCTION | | | |
| Physical | Improves aeration, drainage and moisture retention.<br>"Trash mulch" protects soil from weathering.<br>If too coarse and abundant, may hinder seed preparation. | Creates "crumb structure"—spongy, porous and sticky—that makes an excellent soil conditioner.<br>Dark brown color improves heat retention by soil. | Same as effective humus. |
| Chemical | Provides some soluble nutrients, especially from manures.<br>Leaves a reserve supply of nutrients in the soil.<br>Releases much carbon dioxide as it decomposes. | Mobile in soil; readily releases nutrients to plants.<br>Holds nutrient anions in a form available to plants, but safe from leaching.<br>Increases cation exchange capacity. | Provides long-term nutrient storage and maintains good cation exchange capacity.<br>Toxic substances (as well as nutrients) can be chelated and prevented from entering the ecosystem. |
| Biological | Provides food for microbial decomposers. However, if too carbonaceous, can over-stimulate microbes and lock up available nitrates. | Provides nutrients to microbes as it decomposes.<br>Releases vitamins, hormones, antibiotics, and other biotic substances. | Provides microbial habitat and evidence of healthy biological activity. |

### Benefits of Humus

- Humus can hold the equivalent of 80 to 90% of its weight in water, so soil rich in humus is more drought resistant.
- Humus is light and fluffy, allowing air to circulate easily, and making soil easy to work.
- The sticky gums secreted by microbes in the process of forming humus hold soil particles together in a desirable crumb structure.
- Humus is extremely effective at holding mineral nutrients safe from being washed away in rain or irrigation water, and in a form readily available to plants. Ample reserves of humus also provide additional plant nutrients in times of need.
- Humus is able, because of its biochemical structure, to moderate excessive acid or alkaline conditions in the soil —a quality known as buffering.
- Many toxic heavy metals can be immobilized by soil humus, and prevented from becoming available to plants or other soil organisms.
- Although the color of humus can vary, it is usually a dark brown or black color, which helps warm up cold soils quickly in the spring.

decompose readily is known as effective or active humus. It consists of a high proportion of simple organic acids (fulvic acids) which will dissolve in either acids or bases. This type of humus is an excellent source of plant nutrients, released as soil organisms break it down further, but of little consequence for soil structure and long-term tilth. This kind of humus is mainly derived from the sugar, starch and protein fraction of organic matter.

Humic acids, which dissolve in bases but not in acids, characterize more stable or passive humus; humins, which are highly insoluble and may be so tightly bound to clay particles that microbes can't penetrate them, are the main constituents of the most stable humus. Because stable humus resists decomposition it does little to add nutrients to the soil system, but it is essential to improving the soil's physical qualities. Carbon-14 dating has revealed that very stable humus complexes may survive unchanged for thousands of years. Stable humus originates from woodier plant residues which contain lots of cellulose and lignin.

The status of soil organic matter and humus is a dynamic one, continually changing through the activities of all the creatures that live there. Ideally, there should be a rough equilibrium among the different kinds of humus at any one time, with the more active fractions predominant when plant nutrient needs are highest, then giving way to more stable forms after harvest or when plants are dormant. Fungi and actinomycetes, which are more abundant than bacterial decomposers under cool, damp conditions, are also more important in the creation of stable humus.

The changes are fastest under optimum conditions for soil biological activity, and fresh supplies of raw organic matter must continually be added to keep the cycles moving. Anything which harms or disrupts one member of the soil community can lead to a form of indigestion of the soil. For example, if large amounts of nitrate fertilizer flood the soil system, the bacteria responsible for converting protein fragments into nitrates will be suppressed, in turn "backing up" the whole organic decomposition process. They will recover after a while, but if this process is repeated year after year, the capacity of that soil to digest fresh organic matter will be seriously damaged.

The process by which organic matter and humus breaks down in the soil is called mineralization. While humus is the product of organic matter mineralization, it too can be mineralized under the right conditions. Organic matter management, discussed in Chapter 4, requires that you understand what conditions speed up or slow down mineralization.

Mineralization is fastest when conditions are perfect for bacteria to reproduce: high aeration, adequate moisture, good pH, and balanced mineral nutrients. Cultivation speeds it up by introducing air; if soil is dry, irrigation will also stimulate mineralization. Increasing soil temperature with dark mulch or row covers, or actually heating the soil in a greenhouse bed, encourages faster release of nutrients to plants.

As is true with fertilizing, it's important to understand the concept of "enough" when you choose to stimulate mineralization. Too quick a release of nutrients from organic matter can cause problems which parallel those of overfertilizing: excess plant nitrate uptake or possible leaching of nutrients into groundwater. It's also important to avoid "burning up" vital stable humus reserves by making sure to add enough organic matter to replenish what is mineralized.

Humus tends to accumulate fastest under conditions unfavorable to mineralization: cool temperatures, low pH, and poor aeration. While to some extent this is desirable, the extreme example of going too far is the case of a peat bog, composed of almost pure humus. The key here is balance: an active, healthy biological population will continually be mineralizing humus at the same time that it is being formed. As you become attuned to the signs of biological activity and health in your soil, as well as the rhythms of growth and rest in your crops, you will develop a better sense of "enough" when it comes to humus formation and decay.

## Physical Factors: Soil Structure & Tilth

Tilth is to soil what health is to people. A soil in good tilth is in good physical condition for supporting soil life. Good tilth also means soil is loose and easy to work, so tools as well as plant roots can readily dig in. Moisture and aeration are the key physi-

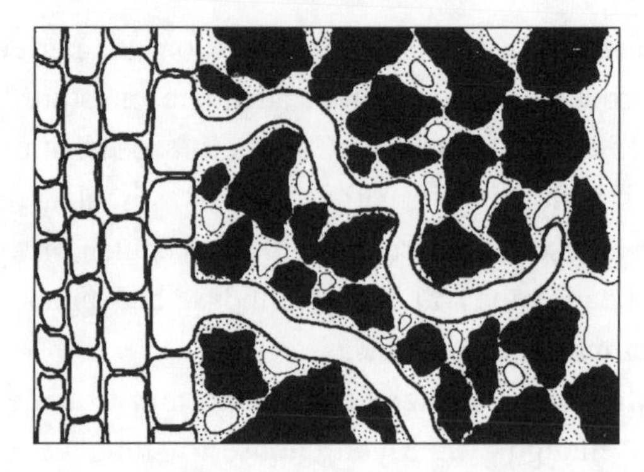

*Figure 1. Soil particles and pore spaces, showing a thin film of water covering each particle. (Drawing by Timothy Rice.)*

cal qualities of soil. The ability of soil to hold water without becoming soggy, and to allow air to penetrate to plant roots and other soil organisms, is vital to every aspect of fertility.

The tilth of your soil is a composite of its texture, structure, aggregation, density, drainage, and water-holding capacity. No matter what kind of soil you start out with, most of these qualities can be improved by increasing its organic matter and humus content.

## Soil Composition

About half the volume of a good, loamy soil is pore space—the area between particles where air and water can penetrate. The pore space is generally an equal volume of air and water, which clings to the surface of soil particles. Don't discount the importance of pore spaces. All the fertilizer in the world won't solve the problems created by dense, compacted soil that is deficient in pore space.

Of the solid half of the soil, about 90% is composed of small bits of the rocks and minerals from which the soil was formed, as well as clays created by the weathering of the parent rock. The remaining 10% is the organic fraction. The influence of this small part of the soil on its ability to support plant growth is tremendous.

The sand and clay components of a soil are largely unalterable—there's not much you can do to change them. But how you manage your soil can have a profound influence on the amount and quality of organic matter it contains. The organic fraction of the soil is a dynamic substance, constantly undergoing change. It consists of living organisms, including plant roots and bacteria, as well as dead plant residues and other wastes. The total weight of the living organisms in the top six inches of an acre of soil can range from 5,000 to as much as 20,000 pounds.

## Fundamental Qualities

Every soil has its own unique physical characteristics, determined by how it was formed. Some of these qualities can be improved with proper management—or made worse by abuse—but others must simply be considered the basic starting point you must work with. Practically, there's nothing you can do to change the depth of bedrock or water table, or to eliminate a steep slope. You can pick rocks out of a very stony soil, but in cold climates, frost heaving will bring more to the surface each spring.

Soil texture is another inherent quality. Texture can range from very fine, mostly clay particles, to coarse and gravelly. Any extreme is undesirable—the ideal loamy texture is a balance of fine clay and silt, combined with coarse sand. The texture of your soil will influence its nutrient status, workability, aeration, and drainage. Clay soils hold water and nutrients well, but can be poorly drained and difficult to work. When they dry out, they form hard clumps, and can take on the consistency of concrete. Sandy soils are generally easy to work and well drained, but have poor nutrient and water holding ability. Very high aeration means organic matter decomposes too rapidly, and little stable humus is formed. (See pages 58–69 for instructions on evaluating your soil's texture and other physical qualities.)

## Structure & Aggregation

Good tilth is less dependent on the composition of your soil than on how it holds together. The ability of soil particles to form stable aggregates, giving it a crumbly, cake-like consistency,

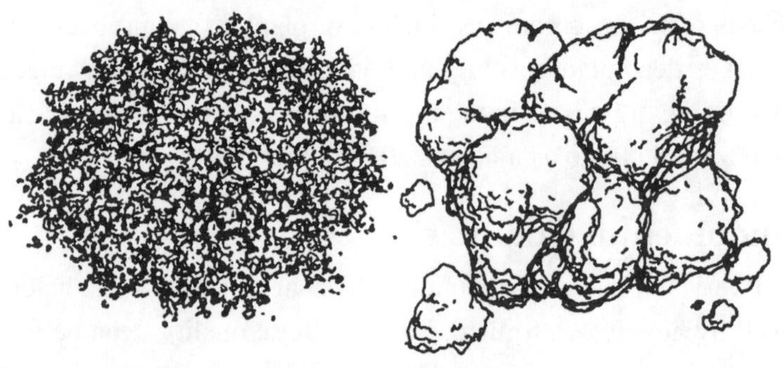

*Figure 2. Comparison of good, crumb-like soil structure (left), with a poor, clod-like structure (right). (Drawing by Stewart Hoyt.)*

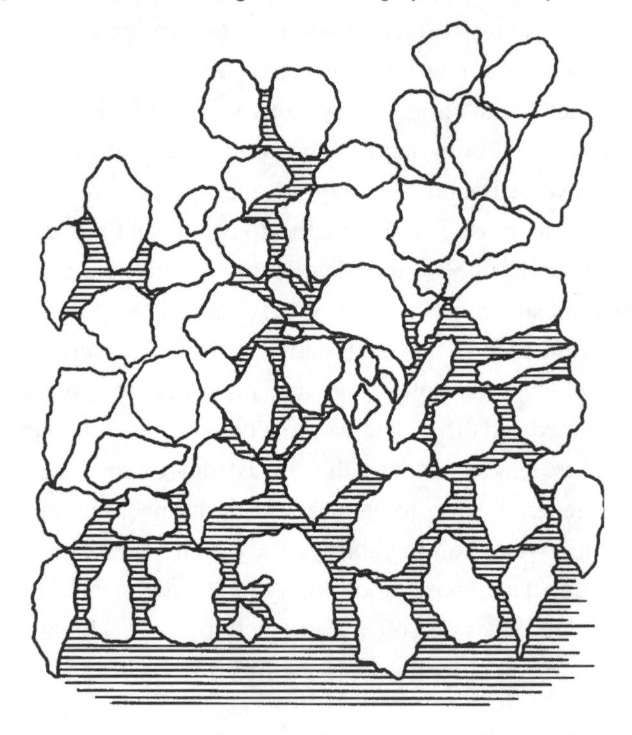

*Figure 3. Moisture moving upward in soil by capillary action. (Drawing by Timothy Rice.)*

*Table 4. Physical properties of soil.*

| Property and definition | Significance in soil | Influence of organic matter |
|---|---|---|
| **Bulk Density:** The weight of unit volume of dry soil, including pore spaces. Expressed as grams per cubic centimeter ($gm/cm^3$). | Indicates how dense the soil is and, therefore, how easily air, water and plant roots can penetrate. (Optimum range: 1.0–1.8 $gm/cm^3$ for compact subsoil). | Increased organic matter leads to decreased bulk density, because organic matter is less dense than soil minerals and gas is released during decomposition. |
| **Pore Space:** The portion occupied by air and water per unit volume of soil. Expressed as a percentage of volume. | Indicates specific aeration and drainage qualities. (Optimum range: 35%–60% for topsoil; 25%–30% for compact subsoil.) | *Increased* organic matter leads to *increased* pore space. Soil organisms also increase pore space by burrowing and eating. |
| **Structure and Aggregation:** Refers to the arrangement of soil particles, their shape, size and stability. | The structure that encourages the most plant growth is *granular:* rounded aggregates that stick together, but shake apart easily. Especially porous granules are called crumbs. | Biological activity is virtually essential for proper granulation. Humus provides a perfect crumb structure that resists compaction. |
| **Oxygen Diffusion Rate:** The rate at which oxygen can be replenished as it is used by respiring organisms. Expressed as grams per cubic centimeter per minute ($gm/cm^3/min$). | Indicates the aeration status of the soil. In addition to pore space for air to enter, there must also be continual diffusion of fresh air into the soil to replace carbon dioxide with oxygen. (Minimum level for root growth is $20 \times 10^{-8}$ $gm/cm^3/min$.) | *Increased* organic matter leads to an *increased* oxygen diffusion rate. Decomposing organic matter (especially plant roots) and mobile soil organisms create air passages in soil. |
| **Field Capacity:** The amount of water held in pore spaces after a fully saturated field has been allowed to drain for 24 hours. Expressed as a percentage of volume. | Indicates the drainage qualities of the soil. A low field capacity means that water runs out too quickly; with a high field capacity, water remains too long in pore spaces. (Well granulated silt loam has a field capacity of about 15%.) | By improving the soil structure, organic matter modulates the field capacity of soils that would otherwise be too wet or too dry. |

determines its structural soundness. The ideal crumb structure is very much a product of biological activity (see Figure 2, page 20). Humus plays a central role in forming soil aggregates, but many soil creatures—most notably earthworms—secrete the sticky gums which are crucial for holding soil particles together. Structure and aggregation can be dramatically improved by increasing humus content and stimulating soil biological activity.

A good crumb structure implies that soil is well aerated, since there will be plenty of pore spaces between the granules. Structure is also essential to the ability of soil to conduct soil moisture upwards toward plant roots. This feature is referred to as capillary action, and works much the same way that oil is taken up into the wick of a lamp. If soil structure is good, the surface may dry out, but moisture will still reach the root zone from deeper soil levels. Adequate soil moisture is crucial not only to replenish what is lost through plant leaves, but also because plant roots take up most of their nutrients dissolved in the thin film of water which coats soil particles.

## Chemical Factors: Nutrient Cycles & Balances

The conventional approach to soil management has been labeled "chemical," in contrast to the "organic" method, which rejects the use of synthetic, petrochemical fertilizers and pesticides. The chemical approach holds that plant roots require certain chemical nutrients, but how these nutrients get there and where they come from matters little. The nutrient elements must be present in a soluble, inorganic form in order for plants to use them.

The ecological viewpoint holds that the effect of fertilizers on soil organisms and the environment is of equal importance to their value as plant food. "Feed the soil, not the plant," and soil organisms will provide a balanced diet to crops. Highly soluble chemicals, though readily taken up by plants, can inhibit or kill soil microbes, and be washed away to pollute groundwater. More-

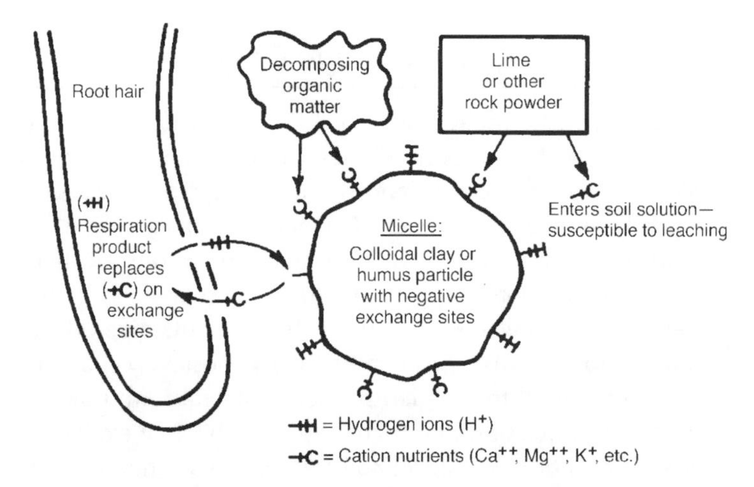


*Figure 4. Cation Exchange Capacity (CEC). The higher the CEC, the more nutrients can be kept available to plants, yet safe from leaching. Cations held on the exchange sites are said to be adsorbed.*

over, plants are also able to absorb and benefit from complex biochemicals such as vitamins and antibiotics, which are not present in artificially synthesized fertilizers.

Public concern over groundwater contamination by nitrate fertilizers, as well as other agri-chemicals, has stimulated greater interest in soil management practices that do not rely on highly soluble materials.

## Soil Chemistry Simplified

Regardless of which approach you adhere to, the fundamentals of soil chemistry are the same. Most chemical interactions—in the soil or anywhere else—take place between particles which carry either a positive or a negative charge when dissolved in water. Positively charged particles are called cations (pronounced "cat-ions"), while anions ("an-ions") carry a negative charge. These particles may be a single element, such as calcium ($Ca_2^+$), or a compound, such as nitrate ($NO_3^-$). The behavior of nutrients in the soil ecosystem is determined by whether they exist as cations or anions.

## Cation Exchange Capacity

Cation Exchange Capacity or CEC is an important measurement of the amount of cation nutrients a given soil is able to store on its clay and humus colloid particles. Colloids have a large number of negatively charged sites on parts of their surface. Positively charged cations are held on these sites, largely protected from leaching away in water, but still available to plant roots. As plants give off hydrogen ions, a waste product that is also positively charged, it is exchanged for needed nutrients like calcium, magnesium and potassium. Nutrients held in colloidal exchange sites may not show up in soil tests because they are not dissolved in water, but are still available to plants through direct contact between roots and soil colloids. This process is referred to as adsorption.

Soils with a high clay and humus content will have the highest CEC, which is measured by how many thousandths of a gram of hydrogen—called milliequivalents—can be held by 100 grams of dry soil. Different kinds of clays have CEC's ranging from 10 to as much as 100, while the CEC of pure humus can approach 200. Very sandy soils will have a CEC of 5 or less.

Think of your soil's CEC as a kind of nutrient savings account. As nutrients are "withdrawn," whether by removing crops or through the prolonged action of water, it is important to replace them to maintain your reserves. These reserves must be well stocked before plants are able to draw on them, so a soil with a high CEC but depleted nutrients will require greater applications of mineral nutrients to restore its fertility than will a similarly depleted but low CEC soil. A high CEC soil with an acid pH will require a larger amount of calcium, in the form of limestone, to correct it than will a low CEC soil with the same pH. Knowing your soil's CEC will help you better understand and interpret your soil test recommendations, as discussed in Chapter 3.

## *Colloids*

One of the most important characteristics of humus is its colloidal nature. Colloids are substances composed of many tiny particles suspended in a gel-like mass, giving them a lot of surface area in proportion to their weight. Protein, which makes up all living cells, is a colloid. Other examples of colloids are milk, mayonnaise, rubber and gelatin. Clay is also a colloid, and the clay component of the soil behaves similarly to humus. Physically, colloids tend to be sticky and absorbent.

Colloids are important chemically because they are covered with negatively charged particles. This makes them able to hold onto positively charged chemical particles, many of which are important soil nutrients. All soil chemical interactions are affected by the soil's clay-humus colloidal content.

## The Basics of Cations

Cation nutrients tend to be metallic mineral elements, important for both plant and microbial nutrition as components of enzymes. They are generally quite water soluble, and enter the soil either through the recycling of organic matter or by addition of mineral nutrient sources such as limestone. Cations are called base elements because they form bases in solution. In humid climates, where there is over 30 inches of precipitation a year, cations tend to become leached out of the topsoil—more slowly if they are stored by soil colloids such as clay and humus. Soils in arid climates, conversely, are usually rich in minerals and so extremely productive when irrigated.

The major cation nutrients include calcium, magnesium and potassium; their functions are summarized in Table 5. Other cation nutrients, needed in minute quantities, are described in the discussion of micronutrients (see page 32).

*Table 5. Major cation and anion nutrients.*

## Major Cation Nutrients

| Nutrient | Natural sources | Forms in soil | Function in plants | Deficiency symptoms |
|---|---|---|---|---|
| Calcium ($Ca^{++}$) | Dolomite; calcite; apatite; calcium feldspars, gypsum. | Most is present as $Ca^{++}$ ion on cation exchange sites, or in soil solution.<br>At high pH, calcium forms insoluble precipitates with phosphorus and some micro-nutrients. | Essential for nitrogen uptake and protein synthesis.<br>Also has a role in enzyme activation and cell reproduction. | Stunted root growth, undeveloped terminal buds, and leaf curl.<br>Pit rot in carrots; blossom-end rot in tomatoes. |
| Magnesium ($Mg^{++}$) | Mica; hornblende; dolomite; serpentine; certain clays. | Present as $Mg^{++}$ ion on cation exchange sites, or in soil solution. | Essential part of chlorophyll molecule.<br>Necessary for phosphorus metabolism and enzyme activation.<br>Often concentrated in seeds. | Yellowing of lower leaves, with venation in green; reduced yields. |
| Potassium ($K^{+}$) | Feldspars; mica; granites; certain clays. | Available as $K^{+}$ on cation exchange sites or in soil solution. (Less than 1% of total soil $K^{+}$ is in available form.) | Essential for carbohydrate metabolism and cell division.<br>Regulates absorption of calcium, sodium and nitrogen. | Weakened stems, scorched leaf edges, necrotic spots, stunted growth, susceptible to disease. |

## Major Anion Nutrients

| Nutrient | Natural sources | Forms in soil | Function in plants | Deficiency symptoms |
|---|---|---|---|---|
| Carbon (C) | Organic matter; respiration of soil organisms. | Organic compounds, carbon dioxide gas in air spaces, and weak carbonic acids. | Basic constituent of all living cells. | If atmospheric carbon dioxide is limited, plant growth is slowed. |
| Nitrogen (N) | Organic matter; atmospheric nitrogen fixed by microbes; small amounts dissolved in rain water. | Organic compounds, nitrites, nitrates, and ammonium (soluble forms). | Basic constituent of protein and genetic material. | Thin stems; yellowing (chlorosis) of leaves, beginning with lower leaves, slowed growth. |
| Phosphorus (P) | Organic matter; mineral powders; some parent materials. | Organic compounds; soluble phosphates; insoluble compounds of iron, aluminum, manganese, magnesium, and calcium. | Essential for genetic material, membrane formation, and energy transfer. | Purpling of leaves, beginning on undersides; stunted roots; slowed growth. |
| Sulfur (S) | Organic matter; atmospheric sulfur fixed by microbes; pollutants in rain water. | Organic compounds; soluble sulfates, sulfites and sulfides. | Important constituent of proteins and certain vitamins. | Deficiency is hard to detect, but resembles nitrogen deficiency, with yellowing of whole plant. |

## Cations, pH & Fertility

When cation or base nutrients are deficient in soil, it becomes acid. pH is a measure of the acidity or alkalinity of soil, determined by the concentration of hydrogen ions in a water or salt solution. Acidity is indicated by a pH below 7.0, which is neutral; pH values over 7.0 indicate alkalinity.

As hydrogen ions replace the cation nutrients held in soil colloidal reserves, soil pH decreases. The solubility, and thus availability to plants, of most nutrients is highest at a slightly acid pH—around 6.3 to 6.8 is optimum. This is also the most favorable range for the functioning of most soil bacteria, though fungi can tolerate a wider pH range. At low pH—below 5.5—most major nutrients and some micronutrients assume insoluble forms. Phosphorus becomes chemically immobilized at both low and high pH, requiring a range between 6.0 and 7.0 for maximum availability.

Many cation micronutrients, including iron, manganese, zinc, copper, and cobalt, become more soluble at low pH but are unavailable under alkaline conditions. In some cases, acid conditions can induce toxicity of these elements. This is also true of certain heavy metals, most notably aluminum, which is naturally present in most soils, and lead, which can sometimes be a contaminant. Neutralizing the pH also neutralizes the heavy metal hazard.

When correcting soil acidity, the object is not so much to neutralize pH as it is to replenish the appropriate cation nutrients—usually calcium and sometimes magnesium in the form of limestone. Applying other alkaline materials, such as sodium bicarbonate, may neutralize the pH, but won't improve soil fertility.

"Acid soil syndrome" is a common problem in areas of high precipitation, where soluble soil bases tend to leach out into the subsoil. Some of the problems associated with acid soil include:

- Interference with the availability of nutrients to plants.

- Increased solubility of iron, manganese and especially aluminum to undesirable levels.
- Reduced bacterial activity, especially of nitrogen-fixing *Rhizobia,* & slower release of nutrients in organic matter.
- Lower total CEC, which further increases nutrients' leachability.

Alkaline soils can be even more difficult to correct. The addition of acid-forming minerals like sulfur is more expensive and temporary than the addition of limestone to acid soils. In many places where soil is naturally alkaline, improper irrigation practices may cause salts to build up in the surface layer, a condition known as salinization. This happens when nutrient-rich water rises to the surface through capillary action, and then evaporates, leaving its minerals behind. Some soils in arid areas are naturally saline. It is difficult and costly to reverse these effects, generally done by leaching the area with large amounts of fresh water—a resource usually in short supply in affected regions.

Among the problems associated with alkaline soils are:

- Unavailability of many nutrients, especially most micronutrients.
- Saline seep, causing soil crusting.
- Toxic levels of sodium, selenium, and other minerals.
- Chemical destruction of organic matter.

## Soil Anions & Their Cycles

Anion nutrients differ from cations in that they are not stored chemically by soil colloids, and form acids in solution. Reserves of anion nutrients are held in the organic portion of the soil, and are released to plants through the decay of organic matter or through the air and water. Depending on the status of soil organisms and the decay cycle, soil anions continually change in form and quantity. As the major building blocks of proteins and carbohydrates, anions are required in larger quantities than are cation nutrients. It is helpful to think of anions as large, soft and changeable in form, while cations are small, hard and durable.

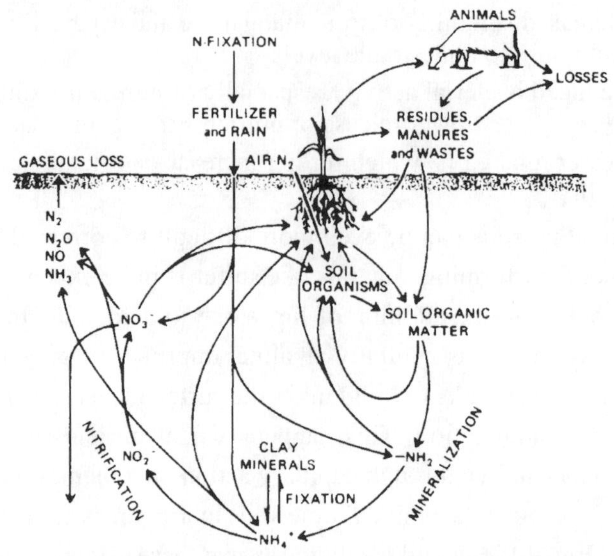


*Figure 5. The Nitrogen Cycle. Plants cannot utilize nitrogen in its gaseous form. In order to pass from atmosphere to plant (and then to animals and people), nitrogen must first be fixed by soil microorganisms. To make synthetic fertilizers, atmospheric nitrogen is artificially fixed through use of huge quantities of natural gas. (33,000 to 40,000 cubic feet of natural gas is required to produce one ton of ammonia.) (Reprinted with permission of MacMillan Publishing Co., from* The Nature & Properties of Soils, *8th ed., by Nyle C. Brady.)*

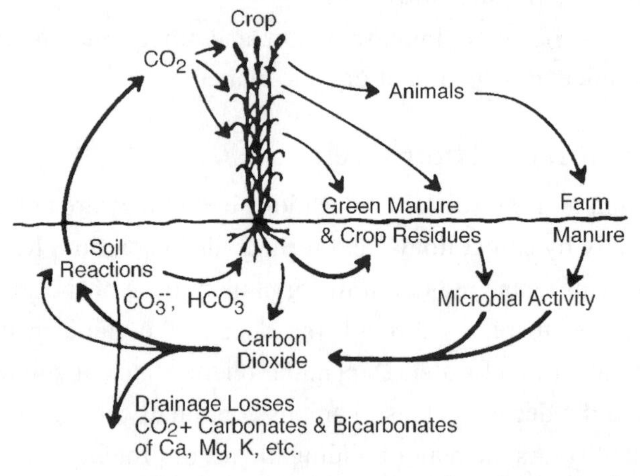


*Figure 6. The Carbon Cycle. Gaseous carbon dioxide is transformed by plants into living tissue. Carbohydrates, proteins, and fats are decomposed by soil organisms, thereby replenishing the supply of atmospheric carbon dioxide. (Reprinted with permission of MacMillan Publishing Co., from* The Nature & Properties of Soils, *8th ed., by Nyle C. Brady.)*

When added to the soil as soluble fertilizers, anion nutrients may be lost because they volatilize into the atmosphere, leach away, or revert to more stable, insoluble forms. These soluble fertilizers may be acid-forming, or otherwise harmful to soil organisms. When substituted for nutrient sources rich in organic matter, they can be likened to an addiction: Higher doses will be required to replace the nutrients that were previously supplied naturally by the soil's ecosystem.

- *Nitrogen* tends naturally towards the gaseous state as its most stable and plentiful form. The nitrate form, in which it is present in the soil solution, is extremely transitory and will fluctuate significantly from day to day and even at different times of day. Although plants cannot use atmospheric nitrogen directly, certain soil microbes, most notably the *Rhizobia* bacteria which live on the roots of legumes, are able to capture it from the air and transform it into a biologically useful form. Free living soil bacteria such as azotobacter and clostridia are even more important for nitrogen fixation. Still other bacteria transform the nitrogen from ammonium to nitrite and then to nitrate form; each step in the natural nitrogen cycle is essential for plant nutrition.
- *Carbon,* the major constituent of plant (and animal) tissue, is more truly the "food" consumed by plants than any mineral. Although abundant in the organic fraction of the soil, carbon is taken in by plants almost entirely from the atmosphere, as carbon dioxide. The carbon dioxide concentration close to the ground can be substantially enriched by the presence of actively decaying organic matter, which directly stimulates plant growth. Lack of carbon dioxide can be a problem in greenhouses, where air circulation must be artificially maintained. An indoor compost pile can add carbon dioxide—and extra heat—to the greenhouse environment.
- *Phosphorus* undergoes some of the most complex chemical interactions of all the elements of soil fertility. It is easily immobilized in the soil through its tendency to form insoluble compounds with calcium and other minerals. At a pH near neutral, even highly soluble phosphorus fertilizers will soon revert to forms indistinguishable from the "natural" rock powders from which they were manufactured. Once added to the soil, immobilized phosphorus stays there a long

time; many soils, especially in the Midwest, have phosphorus reserves created by government encouragement of excessive fertilizer applications. Phosphorus is most readily available to plants when released gradually through the decomposition of organic matter. Its relative immobility means that distribution of phosphorus throughout the soil is only accomplished through the movement of earthworms and other soil organisms. Otherwise insoluble soil phosphorus reserves can become available to plants through the activities of soil microbes.

- *Sulfur,* an essential component of protein and fats, acts much like nitrogen in the soil ecosystem and is particularly important for nitrogen-fixing microorganisms. A special group of microbes transforms organic sulfur into the sulfate form utilized by plants. Sulfur deficiency is rarely a problem, especially where adequate soil organic matter levels are maintained. Air pollution also has had unintended beneficial effects on the sulfur content of soils downwind. Deficiencies have risen with increased use of highly concentrated phosphate and nitrate fertilizers, which lack the sulfur impurities found in the lower grades. Sulfur is often needed as a nutrient and an acidifier for alkaline soils.

## Micronutrients

Micronutrients are elements which are important in very small amounts for the proper functioning of biological systems. Sometimes called "trace elements," over a dozen of them have been identified as essential in minute quantities for plant, animal or microbial enzyme functions. Most of the important micronutrients, such as iron, zinc, copper, and manganese, are cations; boron and molybdenum are the most important anion micronutrients.

Micronutrients occur, in cells as well as in soil, as part of large, complex organic molecules in chelated form. The word "chelate" (pronounced "KEE-late") comes from the Greek word for "claw," which indicates how a single nutrient ion is held in the center of the larger molecule. The finely balanced interactions between micronutrients are complex and not fully understood. We do know that balance is crucial; any micronutrient,

*Table 6. Some important micronutrients*

| Nutrient | Desired amount in soil (ppm)* | Function in plants | Deficiency symptoms |
|---|---|---|---|
| Iron (Fe) | 25,000 | Chlorophyll synthesis; oxidation; constituent of various enzymes and proteins. | Chlorosis (yellowing), larger veins remain green; short and slender stems. |
| Manganese (Mn) | 2,500 | Synthesis of chlorophyll and several vitamins; carbohydrate and nitrogen metabolism. | Yellow mottling of leaves; pale overall coloring; poor maturation and keeping quality. |
| Zinc (Zn) | 100 | Formation of growth hormones; protein synthesis; seed and grain production and maturation. | Late summer mottling of leaves. Early defoliation of fruit trees. |
| Copper (Cu) | 50 | Catalyst for enzyme and chlorophyll synthesis, respiration, carbohydrate and protein metabolism. | Yellowing and elongation of leaves. Onions are soft, with pale yellow scales. |
| Boron (B) | 50 | Protein synthesis; starch and sugar transport; root development; fruit and seed formation; water uptake and transport. | Growing tips die back. Heart rot of root crops; corky core of apples; poor legume nitrogen fixation. |
| Molybdenum (Mo) | 2 | Essential for symbiotic nitrogen fixation and protein synthesis. | Pale, distorted, narrow leaves, leaf roll; poor nitrogen fixation. |

**Approximate values indicate relative proportions.*

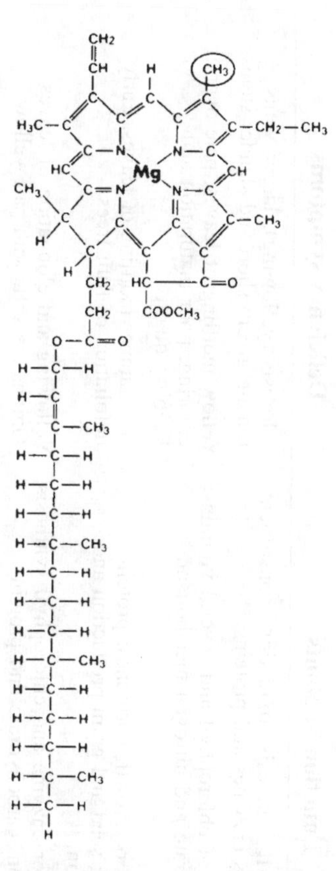


*Figure 7. This representation of chlorophyll illustrates chelation. Note the single atom of magnesium, which occupies a central position in this organic complex. Metallic elements (cations) bound into organic molecules in this fashion perform many functions in living systems.*

when present in excessive amounts, will become a poison, and certain poisonous elements, such as chlorine, are also essential micronutrients.

For this reason natural, organic sources of micronutrients are the best means of supplying them to the soil—they are present in balanced quantities and not liable to be overapplied through error or ignorance. When used in naturally chelated form, excess micronutrients will be locked up and prevented from disrupting soil balances. Soil humus reserves also serve to chelate excess metals—nutrients as well as toxins. Unless a specific micronutrient deficiency has been diagnosed by a soil test, the best way to provide adequate supplies is by building organic matter and applying balanced sources of minerals such as rock powders and seaweed. (See Chapter 4 for more information on micronutrient fertilizers.)

## A Balancing Act

Balance is the crucial concept for understanding the relationship between chemical nutrients and soil fertility. Fertility requires not only sufficient quantities of nutrients, but their presence in balanced form. In many cases, too much of one nutrient will lock up or interfere with the absorption of another. Phosphorus is the classic example; it will become immobilized at low pH by high concentrations of aluminum, zinc and iron, and at high pH by too much calcium. Potassium and magnesium will each interfere with the availability of the other, when present in excess. In the case of carbon and nitrogen, too much carbon will stimulate soil microbes to grow and take up all the available soil nitrogen, resulting in a temporary deficiency until the microbes die and release their nutrients to the soil system.

Cation balances have received a lot of scientific attention. Base saturation refers to the percentage of a soil's CEC (see sidebar, page 24) occupied by bases—cations other than hydrogen or aluminum.

Some efforts have been made to find the "ideal" cation balance or base saturation ratio. One very influential scientist was Dr. William Albrecht, who conducted research at the University of Missouri in the 1940s. Albrecht's key contribution was to point out the importance of calcium as a major ingredient of fertility, contending that it was the calcium in limestone, not its acid-neutralizing ability, that made it an important fertilizer. He also developed a formula for an optimum base saturation ratio, emphasizing calcium, which has been used by many soil labs to eval-

*Table 7. Nutrient interactions.*

| Nutrient | Deficiency may be induced by excess of: | Excess may induce deficiency of: |
|---|---|---|
| CATIONS | | |
| Calcium | Aluminum | Magnesium, potassium, iron, manganese, zinc, phosphorus, boron |
| Magnesium | Calcium, potassium, ammonium | Potassium, zinc, boron, manganese |
| Potassium | Magnesium, calcium, ammonium | Magnesium, boron |
| Iron | Phosphorus (high pH), manganese (low pH), calcium, copper, aluminum, zinc | Zinc, manganese |
| Manganese | Iron, copper, zinc, calcium, magnesium | Iron |
| Zinc | Phosphorus, nitrogen, magnesium, iron, copper, calcium, aluminum | Iron, copper, manganese |
| Copper | Phosphorus, zinc, nitrogen | Iron, zinc, manganese |
| ANIONS | | |
| Carbon | Sulfur, nitrogen, phosphorus | Sulfur, nitrogen, phosphorus |
| Nitrogen | Carbon, phosphorus | Phosphorus |
| Phosphorus | Calcium, nitrogen, iron, aluminum, manganese | Zinc, copper, nitrogen |
| Sulfur | Carbon, nitrogen | |
| Boron | Calcium, potassium | |

uate mineral balances. While it is a useful guideline, Albrecht's ratio is not universally accurate, and should not be relied on exclusively to determine fertilizer needs.

The ideal proportion of anion nutrients is the balance that is normally found in humus: 100 parts carbon:10 parts nitrogen: 1 part phosphorus:1 part sulfur. The importance of the carbon to nitrogen ratio was described earlier; the ratio of nitrogen to phosphorus is also important to proper plant nutrition, since inadequate nitrogen slows the growth of roots and therefore their ability to reach phosphorus supplies.

Micronutrient problems are as often a result of imbalances as of absolute deficiencies. New information is continually being discovered about previously unknown interactions between major and minor nutrients in the soil ecosystem. This is why the "cookbook" approach to soil chemistry can get you into trouble; the best nutrient sources are those which are naturally balanced. Table 7 gives some indication of the complexity of known nutrient interactions.

## Biological Factors: Life In & On the Soil

### The Soil Community

The cycles that permit nutrients to flow from soil to plant are all interdependent, and proceed only with the help of the living organisms that constitute the soil community. Soil microorganisms are the essential link between mineral reserves and plant growth. Animals and people are also part of this community. Unless their wastes are returned to the soil—for the benefit of the organisms that live there—the whole life-supporting process will be undermined.

Soil organisms—from bacteria and fungi to protozoans and nematodes, on up to mites, springtails and earthworms—perform a vast array of fertility maintenance tasks. Ecological soil management aims at assisting them—not substituting a simplified chemical system to provide nutrients to plants. Once disrupted by excessive use of soluble fertilizers, the restoration of a healthy soil ecosystem can be a long and costly process.

### Soil Inhabitants

Microscopic plants and animals form the basis for the soil food web. Most contribute directly to humus formation and the release of nutrients from organic matter. Stable humus, in fact, consists largely of microbial remains. In cool, humid climates, fungi and molds are more significant than bacteria for humus development. Beyond their importance for soil health, these microbial decomposers are essential to all life on earth, since they are responsible for virtually all organic waste recycling.

Other creatures, both microscopic and visible, make important contributions to soil health, most notably the earthworm. Plant roots themselves are major participants in the soil ecosystem, and significantly affect the environment which sustains them. Once you become aware of the astonishing number and variety of life forms which are constantly growing, reproducing and dying in every crumb of soil—billions in each gram of healthy topsoil—it is impossible to pick up a handful without a sense of awe. Each organism has a role to play in the soil ecosystem:

- Producers create carbohydrates and proteins from simple nutrient elements, almost always by capturing energy from sunlight through photosynthesis. Green plants, including blue-green algae, are the producers of the soil. A few specialized bacteria, known as autotrophs, are also able to synthesize their own food from carbon dioxide and mineral elements in the soil.
- Consumers are just about everyone else, all of whom depend on the food created by green plants for their nourishment. Primary consumers eat plants directly, while secondary and tertiary consumers feed on other consumers. All animal life, from simple protozoans to humans, as well as non-photosynthesizing plants such as yeasts and certain other fungi, fall into this category.
- Decomposers perform the critical function of bringing the basic chemical nutrients full circle, from consumers back to producers. They are all bacteria or fungi, and are found almost exclusively in the soil; about 60 to 80 percent of the total soil metabolism is accounted for by microbial decomposers. Without them, life would grind to a halt as we suffocated in our own wastes.

## What They Need

If the surest route to improving soil fertility is to provide the most hospitable conditions for soil life, understanding the basic needs of soil organisms is the first step. Successful composting requires the same knowledge: Provide soil organisms with adequate food, air and water, and, depending on other environmental factors such as temperature, they will flourish.

Both soil and compost creatures need the same food: raw organic matter with a balanced ratio of carbon to nitrogen—approximately 25-30 to 1. Carbon, in the form of carbohydrates, is really the main course for soil organisms—they will grow quickly given lots of it, and scavenge every scrap of nitrogen from the soil system to go with it. That's why adding lots of high-carbon materials to your soil can cause nitrogen deficiencies in plants. In the long term, though, carbon is the ultimate fuel for all soil biological activity, and therefore of humus formation and productivity. A balanced supply of mineral nutrients is also essential for soil organisms, and micronutrients are key to the many bacterial enzymes involved in their biochemical transformations. Balanced nutrients provide a favorable pH, though different organisms are more sensitive to acid or alkaline conditions.

Air is also crucial for soil health, although certain bacteria can live without it. In fact, much effort in soil management is directed toward improving soil aeration—no amount of fertilizing can compensate for lack of air, since plant roots cannot take full advantage of available nutrients if they are suffocating.

Water is also strictly essential, but not to the extent that it drives out air. The ideal biological environment consists of a thin film of moisture clinging to each soil particle, with lots of air circulating between them. Rain and irrigation, of course, play a central role in adding needed soil moisture, but good structure is required to conduct moisture upward from reserves in lower soil strata (see page 20).

Although temperature has critical effects on biological activity, every specific soil community has evolved to accommodate the natural climate variations in its environment. Your only role in adjusting the temperature might be to moderate severe winter cold or summer heat by mulching, or to heat up small areas with season extension devices.

## Soil Superstars

Despite the many volumes which have been written about soil biology, knowledge of the kinds of organisms which live in soil and how they interact is extremely limited. Although some scientists have tried to work out biological assays to identify a soil's characteristics and needs by examining its living population, such tests are still extremely complex to carry out and difficult to interpret accurately.

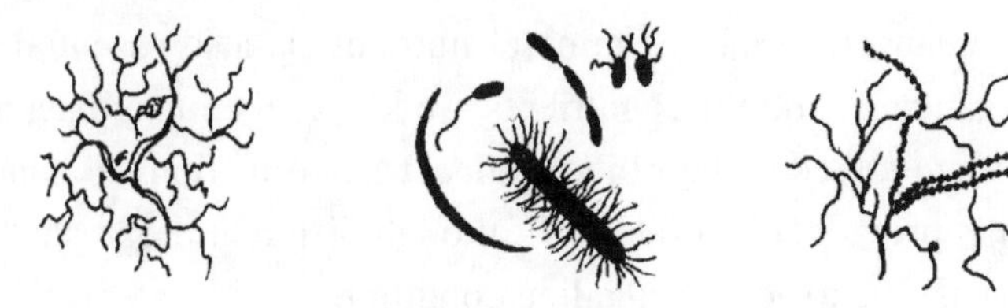

The three most important plant microorganisms of the soil. Fungal mycelium (left), various types of bacteria cells (center) and actinomycetes threads (right). The bacteria and actinomycetes are much more highly magnified than the fungus.

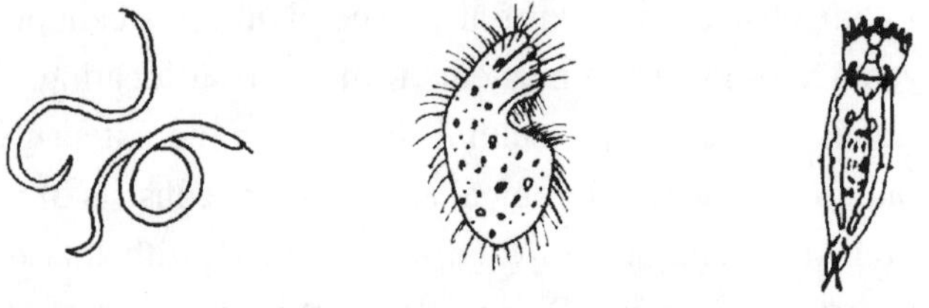

Parasitic nematodes (left), a ciliated protozoan (center), and a common rotifer (right).

Some soil organisms especially important in the nitrogen cycle. (Left to right) *Azotobacter*, nitrate bacteria, and nodule organisms of alfalfa.

*Figure 8. Important Soil Organisms. (Drawing by Timothy Rice.)*

## *Bacteria & Breathing*

The soil micro-universe is divided into two types of bacteria: those which need air and those which don't. The availability of air thus determines which kinds will flourish, and how vigorously they will grow.

- Aerobes require air in order to live. The bacteria which mediate the soil nitrogen and sulfur cycles, as well as many important decomposers, are aerobes. Other types of soil organisms are always aerobic, including plant roots. Some bacteria can survive in either aerobic or anaerobic conditions, but will only grow and thrive if they have air.
- Anaerobes, which are all bacteria, can live happily without air, and in fact may be killed if exposed to it. The organisms responsible for diseases like botulism and tetanus are famous examples. There are many anaerobic decomposers, which often generate some foul-smelling waste products, such as hydrogen sulfide, and the common term for the process of anaerobic decomposition is putrefaction. Anaerobic bacteria can also generate useful by-products such as methane gas, which is sometimes used as an energy source.

Table 8 summarizes the major types of soil life-forms. A few portraits of the more familiar and celebrated soil inhabitants, in order of size, follows:

**Bacteria:** Bacteria are the most numerous and varied of soil organisms, ranging from a few hundred million to 3 billion in every gram of soil. Under the right conditions they can double their population every hour. The top 6-8 inches of soil may contain anywhere from a couple of hundred pounds to 2 tons of live bacteria per acre.

Bacteria vary in their requirement for air, but most beneficial ones need it (see above). If enough moisture and food are present, bacteria do best at temperatures of 70-100° F, and at pH close to neutral. Adequate calcium is crucial, as is a balance of micronutrients, which are essential to the enzymes employed by bacteria

*Table 8. Soil organisms.*

| Organism | Approximate soil population | Special requirements | Source of nutrition | Role in ecosystem |
|---|---|---|---|---|
| MICROFLORA | | | | |
| Fungi: yeast, molds, mycorrhizae. | $10^5$–$10^6$ per gram | Will tolerate wide pH & temperature ranges. | Organic matter or nutrients from plant roots. | Humus formation; aggregate stabilization. Create antibiotics; cause plant diseases; make phosphorus available. |
| Actinomycetes. | $10^7$–$10^8$ per gram | Need aeration, moisture, and pH 6.0–7.5 | Organic matter. | Humus formation. |
| Bacteria autotrophs: *Azotobacter, Rhizobia Nitrobacter*. Heterotrophs: decay organisms. | $10^8$–$10^9$ per gram | Most need air (aerobes), and exchangeable calcium. Temperature of 70°–100° F; pH 6–8. | Autotrophs consume simple nutrients from soil & air. Heterotrophs break down organic matter. | Autotrophs are nitrogen-fixers, sulfur oxidizers, nitrifiers. Some cause disease. Heterotrophs are decomposers. |
| Algae: Green, blue-green. | $10^4$–$10^5$ per gram | | Photosynthesis. | Add organic matter to soil. Some fix nitrogen. |
| MICROFAUNA | | | | |
| Nematodes. | 10–100 per gram | | Organic matter, other microbes, plant roots. | Secondary consumers. Some are pests and parasites. |
| Protozoa, rotifers. | $10^4$–$10^5$ per gram | | | |

| Organism | Approximate soil population | Special requirements | Source of nutrition | Role in ecosystem |
|---|---|---|---|---|
| INSECTS & MOLLUSCS | | | | |
| Mites, springtails, spiders, sowbugs, ants, beetles, centipedes, millipedes, slugs, snails. | $10^3$–$10^5$ per $m^2$ | | Microflora, microfauna, other insects, plant roots and residues, nematodes, molluscs, detritus (waste matter), organic matter, weak plants. | Aerate and mix soil. Leave remains for decomposers. Cull weak or diseased plants. |
| EARTHWORMS | 30–300 per $m^2$ | | Raw organic matter. | Aerate and mix soil. Leave nutrient-rich casts. |
| MAMMALS | | | | |
| Moles, mice, goundhogs. | Variable | | Earthworms, insects, molluscs. | Mix and pulverize soil. Leave wastes. |
| PLANT ROOTS | 100–6,000 lbs. per acre | | Photosynthesis; nutrient ions and molecules. | Remove water and nutrients; leave residues and exudates. |

## *Plant-Microbe Symbiosis*

The mutually beneficial relationships between plant roots and soil microbes are complex and widely varied. Some of these arrangements are well understood, but many remain mysterious. As we learn more about the workings of these microorganisms, they will undoubtedly be used more widely to improve crop production.

A few bacteria are able to convert nitrogen gas from the air into a form usable by the roots of the plant with which they associate. *Rhizobia* bacteria, for example, form visible nodules on the roots of legumes, whose growth is greatly enhanced by the nitrogen fixed there. When the legumes are returned to the soil, as with green manure crops, the nitrogen fixed by the *Rhizobia* becomes available to the subsequent crop (see Table 15 B). When a legume is grown in association with another crop, for example grasses or grains, the nitrogen fixed by the *Rhizobia* is available to the associated crop while the two are growing together. Because not all soils contain these desirable nitrogen-fixing bacteria, when farmers plant legumes they often inoculate either their soil or seed with preparations containing *Rhizobia.*

Other soil organisms, such as the actinomycete *Frankia,* the bacteria *Azotobacter* and some blue-green algae, are likewise capable of fixing nitrogen, with and without host species.

It has been estimated that more than 80% of plants have symbiotic associations with fungal mycorrhizae, whose pervasive filaments extend the reach of plant roots in the soil, often improving the roots' ability to absorb nutrients. Some of these mycorrhizae are relatively nonspecific, able to penetrate the roots of many different species of plants in various ecosystems. Others are quite specific, surrounding the roots of only one plant species in one type of soil.

to perform their crucial biochemical tasks. Unfavorable conditions rarely kill bacteria off completely; they will either stop growing and form spores to wait for better times, or adapt to the changed conditions as genetic mutations quickly spread to new generations. This adaptability can work against you when the organism in question causes a plant disease, though. If any soil nutrient is in limited supply, bacteria will be the first to consume it; plants then must wait to partake until the microbes die and decompose.

Bacteria have a virtual monopoly on three basic soil processes that are vital to higher plants: nitrification, sulfur oxidation, and nitrogen fixation. Nitrifying bacteria transform nitrogen in the form of ammonium, a product of protein decomposition, into nitrate, the form most available to plants. The sulfur oxidation process is analogous to nitrification. Nitrogen-fixing bacteria are able to transform elemental nitrogen from the atmosphere into protein, and eventually make it available to other organisms—a process imitated by humans at a high energy cost. They may live in symbiosis with plant roots, such as the members of the *Rhizobia* family, or they may be free-living soil dwellers, such as *Azotobacter*.

**Fungi:** Yeasts, molds, and mushrooms are all fungi, and only yeasts have little presence in the soil. Although fungi are plants, they do not contain chlorophyll and so must depend on other plants for their nourishment. Molds may be as numerous as bacteria in soil, and will outnumber them under conditions of poor aeration, low temperature, and acidity, which they tolerate more easily than do bacteria. Although many plant diseases are caused by soil-dwelling molds such as *Fusarium* and *Aspergillus*, those in the *Penicillium* family are well-known as disease fighters. Molds are especially important for humus formation, predominating in humus-rich acid forest soils.

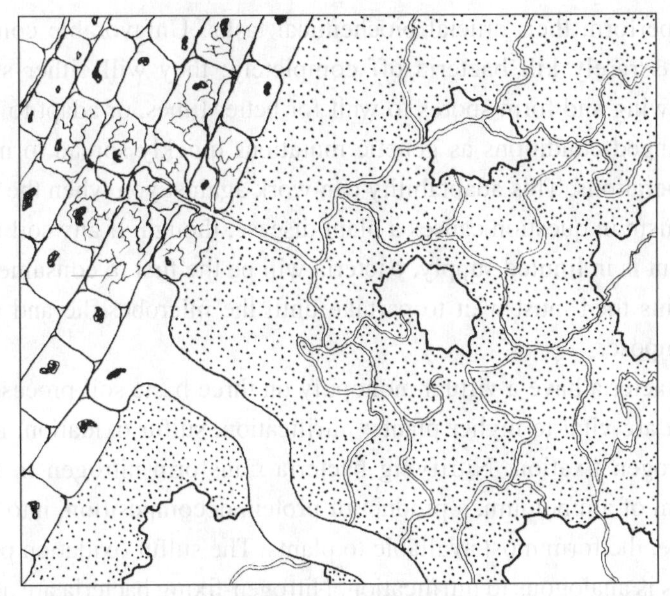

*Figure 9. Mycorrhizae infecting a plant root & extracting nutrients from rock particles. (Drawing by Timothy Rice.)*

One extremely significant group of fungi are called *mycorrhizae*, a term meaning "fungus root." These fungi enter into symbiotic relationships with plant roots of many kinds, and are thought to be essential for the health of trees such as pine and birch. The fungi are able to convert otherwise insoluble nutrients, most notably phosphorus, into biological forms, and in turn receive carbohydrates from their host plants. Many crop plants are known to enter into mycorrhizal associations, but they are most significant for plants growing on poor soils, where the fungal ability to extract nutrients from rock particles is most critical to the host plant's nutrition.

**Actinomycetes:** These microbes are like a cross between bacteria and fungi, and are the most numerous soil organisms after bacteria. The characteristic aroma of freshly plowed earth is attributed to actinomycetes, which play a critical role in organic matter decomposition and humus formation. They need plenty of air and a pH between 6.0 and 7.5, but are more tolerant than ei-

## *The Noble Worm*

A good earthworm population is universally considered to be a sign of healthy soil. Unparalleled as soil excavators, earthworms spend their lives ingesting, grinding, digesting, and excreting soil—as much as 15 tons per acre goes through earthworm bodies in a year. Earthworm castings are richer in nutrients and bacteria than the surrounding soil, and may add up to as much as 8 tons per acre in cultivated fields. Their contribution to drainage and aeration, soil aggregation, and transport of nutrients from the subsoil is significant as well. It is for good reason that Charles Darwin extolled earthworms as the "intestines of the soil."

Of about 200 known earthworm species, *Lumbricus terrestris* is the most common—interestingly enough, it is not native to North America, but came with the Europeans and turned out to be better adapted to cultivated conditions than its native predecessor. Earthworms, unlike the types used for composting, prefer cool temperatures—about 50° F. is optimum. They need good aeration and enough but not too much moisture. Although some species can tolerate fairly acid soils, most require adequate calcium supplies and thus more neutral pH. They are also sensitive to many toxic pest and weed control chemicals, as well as fertilizers with a high salt index.

ther bacteria or fungi of dry conditions. Their intolerance of low pH can be used to advantage in preventing potato scab, a disease caused by an actinomycete. Manure is especially rich in actinomycetes, which is why many people consider manure to be essential for making high quality compost.

**Algae:** Algae are single-celled plants, usually containing chlorophyll, and are slightly less numerous in the soil than are fungi. Blue-green algae are common in many kinds of soils, but are particularly important in paddy rice culture because of their

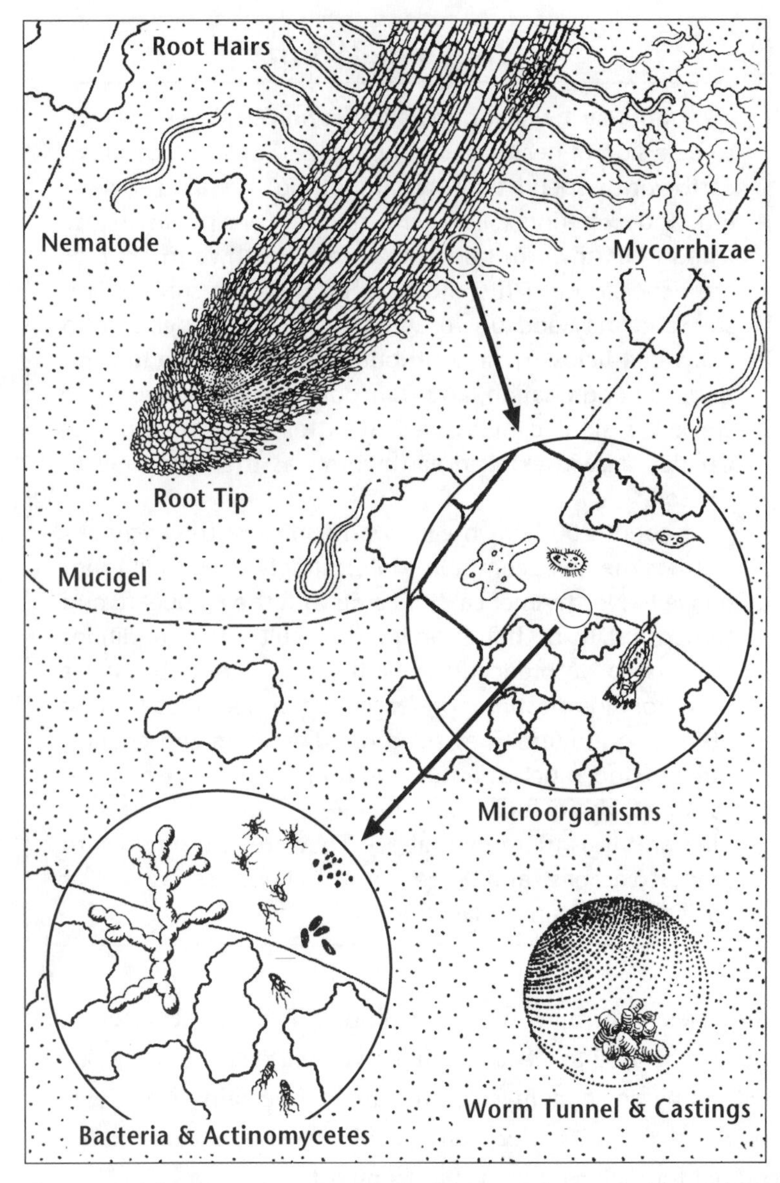


*Figure 10. The rhizosphere, a region approximately 3 mm wide, and the zone of highest biological and chemical activity where soil, root and microbes interact. Mucigel is a gelatinous substance surrounding the root which is both the mediator and the product of the root-soil-microbe community. (Drawing by Timothy Rice.)*

tolerance for high moisture levels and their ability to fix atmospheric nitrogen. All algae growth is greatly stimulated by farm manure.

**Microfauna:** Microscopic soil animals include nematodes, protozoa and rotifers. Nematodes, commonly called threadworms or eelworms, are extremely widespread and numerous in most soils. Although they are often thought of as troublesome plant pests, the most common kinds help break down organic matter or prey on bacteria, algae, or other soil animals. Some parasitic nematodes are used as biological control agents for soil-dwelling pests such as cabbage root maggots. Protozoans are one-celled animals which are larger than bacteria, often using them as a food supply. Rotifers are common in wet soils, feeding by spinning around and sweeping food particles into their "mouths."

**Earthworms & Other Macrofauna:** Of the numerous animals who make their homes in the soil, from miniscule spiders to prairie dogs, the earthworm is the most closely identified with soil health. Other small soil animals include mites and spiders, beetles, springtails, flies, termites, ants, centipedes, and slugs, as well as the larval forms of many butterflies and moths. Many play an important role in breaking down organic materials into smaller pieces and simpler compounds; some are significant as plant pests. Mammals such as moles, voles, gophers, and other burrowers may sometimes be a nuisance when they decide your broccoli looks good to them; however, they too contribute to the soil ecosystem by keeping pest populations in check, mixing soil, and depositing droppings.

**Farm Animals:** The importance of farm animals to soil fertility sometimes gets overlooked. Although modern "factory farming" concentrates too many animals in small areas, the ideal farming system should include some animals. Besides contributing valuable manure, farm animals are important for a number of ecological soil-building practices:

- Soil-building rotations require sod crops and legumes, which are often only economically feasible when used as animal feed.
- Animals are often the only means to harvest a crop (besides trees) on soils that are too wet, steeply sloping or stony to cultivate.
- Pigs and chickens can improve the soil by acting as living rototillers, scratching and aerating a patch before crops are planted. They also eat weeds and ground-dwelling pests.

## Life in the Root Zone

Plant roots themselves play an important role in soil ecology. The largest numbers and kinds of organisms are found in the uppermost layers of the soil, closer to fresh sources of air, water and food. Some biological activity happens even at fairly deep levels, especially where earthworms and other animals burrow, and deep-rooted plants grow. However, in the area immediately surrounding plant roots, known as the rhizosphere, are found concentrations of 10 to as many as 100 times more organisms than elsewhere in the soil. A soil such as that found under permanent grass sod, totally permeated by fibrous masses of roots, will inevitably have a healthier, more robust microbial population than one with cleanly cultivated row crops.

Most of the important soil biological transformations take place in the rhizosphere, especially nitrogen fixation and mycorrhizal associations. The outer coating of the growing root tip, called the mucigel, is a fascinating substance which is a product of both the root and the micro-community around it. A gelatinous substance secreted by the root, it is a rich mass of microbes and chemical nutrients which connects the plant directly to the life of the soil.

Some ways in which plant roots interact with the rest of the soil community include:

- Roots can take up cation nutrients directly from exchange sites on soil particles, in exchange for hydrogen, through a process known as adsorption (see sidebar, page 24).

- Plant roots give off carbon dioxide in the process of respiration, just as animals do. Together with the release of hydrogen described above, this creates slightly more acid conditions in the rhizosphere, since carbon dioxide forms a weak acid when dissolved in water. This slight acidity helps to make otherwise insoluble phosphorus and micronutrients more available.
- Roots give off certain biochemical compounds called exudates, which sometimes act as phytotoxins, chemical inhibitors of competing plant species—a process called allelopathy. Winter rye, for example, gives off exudates which suppress couchgrass growth.
- The dead tissue continually sloughed off by growing roots is excellent food for microorganisms. The organic contribution of the root portion of a green manure crop often is more substantial than the part you see above the surface.
- Plant roots are able to take up many complex organic compounds such as hormones, vitamins, antibiotics, and humus fractions, as well as toxic substances like pesticides and herbicides. This is an important counter to the argument that the source of plant nutrients—whether chemical fertilizers or compost—is irrelevant to plant health.

## Soil & Civilization

All land-dwelling animals, including humans, are members of the soil community. Human societies disregard this fact at their own peril—soil fertility has historically been squandered for the immediate enrichment of a few at the expense of future generations. Cultural values—ethics, aesthetics and spiritual beliefs—have a profound influence on how soil is treated.

Not only the farm itself, but also the society of which it is a part must be viewed as components of the soil ecosystem, for all support and maintain one another and none can exist independently. Without a good sized non-farm community nearby, for example, marketing becomes a problem for the farmer; too large a non-farm community exerts pressure to convert productive farmland to other uses. A whole book could be written about the effects of political and economic pressures on soil fertility—especially in the "Third World," where peasants are forced to produce export crops for foreign exchange instead of food for their families.

The 1992 United Nations Earth Summit acknowledged the importance of sustainable agriculture as a means of reversing worldwide environmental degradation. Implementing its recommendations will require widespread public consciousness raising. Political and social activism are, more clearly than ever, essential components of soil stewardship.

# 3

# Observing & Evaluating Your Soil

Conventional agriculture has frequently confused means and ends, establishing "goals" such as a 150 bushel corn yield, then testing soil nutrient levels for comparison with levels "known" to be appropriate to the desired yield. Commonly, both the target corn yield and the associated target soil nutrient levels are considered in complete isolation from the broader ecological, economic and family situation of which they are a part. Blatant conflict of interest is far from rare, as it is often the fertilizer company doing the sampling and making the recommendations as a "service" to the farmer.

In ecological agriculture, by way of comparison, the farm family's goal may well be expressed in terms of health, happiness, beauty, and prosperity.

Soil testing, and other tests as well, become tools to evaluate how well the farm system is moving toward the different components of the goal.

## Recordkeeping

To be a good farmer, you must be a keen observer of nature —and you must be able to use your observations to improve your management practices. Hence the truism, "The best fertilizer is the footprint of the farmer." You may try out a technique developed at a research station, borrow an idea from a neighbor, or simply follow a hunch. The skill lies in judging how well it works for you.

One of the most important tools of soil management is an accurate set of records. Your farm is unique: what works elsewhere may not work the same way here. To evaluate a new method, you try it under different conditions, keep track of the results, and then compare. This is the essence of the scientific method—what an experienced farmer might call common sense.

Your recordkeeping system should be a good match with the level of complexity of your farm system. You should be able to track the progress of livestock, crop production, processed products, farmstand sales, or orchard management separately. There are many computer-based recordkeeping and farm planning systems available to help you evaluate the financial results of your management decisions. The financial viability of any farm rests to a large extent on your ability to make changes based on information in your records.

Maps are one of the primary farm recordkeeping tools. Aerial photos, used to verify participation in federal conservation programs, are often available from the Soil Conservation Service. You should have an overall map of your farm, showing relative location and size of each field unit, building, waterway, road, and other significant features. Include as much detail as possible about slopes, rocky outcroppings, wet areas, and other characteristics critical to how the land is used. This map should be big enough to note what crop is being grown in each field—keep a master and photocopy it each season so you can look back at a glance to find each field's cropping history.

Some kind of journal is helpful for keeping track of information as it comes up, such as weather (especially rainfall), pest, weed and disease problems, crop growth and development, as well as your management activities—planting and transplanting, cultivating, pruning, spraying foliar nutrients, fertilizing, and harvesting. Refer to your map to indicate where something happened, whenever appropriate.

A diary format works well for much of this recordkeeping, and also allows you to reflect on the possible causes of problems and to make note of ideas for improvement as they come up. Some people use a large wall calendar to record activities, as well as to remind themselves of what needs doing when. A looseleaf notebook, with separate pages for each field or rotation unit, can help you quickly find out when a given area has had a green manure crop or compost application. You can also have a page describing the yearly soil observations you've made, along with copies of your soil test results.

One section of your notebook can be devoted to background information about your farm. Ask your local Soil Conservation Service office for the most recent soil survey for your county (not every county has complete surveys available). This will include extensive information about your soil type and its suitability for different uses. Drainage characteristics, geological information (such as effects of glaciation and type of bedrock) and slope are all described in detail.

Beyond just your land, it helps to know about your own bioregion—the area defined by common ecological factors. Elevation, temperature and precipitation patterns, waterways, and the types of vegetation and wildlife that predominate in your region are all important pieces of information. Where do the prevailing winds come from, and how strong are they? Is there a nearby lake that moderates temperatures? When it is put together, this information can help you make better educated guesses about the sources of problems or the probable success of a new farm crop.

If you are interested in organic certification, getting into a good recordkeeping habit is crucial. Extensive and detailed records are a key requirement of every *bona fide* organic certification program. You need to be able to go back at least three years in your soil management history, and show what fertility improvement and pest control measures you have used. Keep receipts for any purchased fertilizers, soil amendments and other farm inputs as well.

If some of your fields are not managed organically, you must be able to show adequate buffer zones and facilities to prevent commingling of organic and conventional products. This includes a system for tracking the life cycle of everything you sell from your farm, how it was produced, and where it went. Finally, your records will enable you to develop a credible farm plan, a document central to the provisions of the Federal law. For more information on organic certification requirements, consult one of the organizations listed in the Appendix.

## Monitoring Tilth: Reading The Field

### Weeds

There can be wide variations within one field: differences in terrain and existing vegetation, for example. Checking for soggy areas in spring or after a good rain will reveal spot drainage problems. The condition of existing vegetation—whether it's weeds, forage, or field crops—can provide many clues about the status of your soil. Weeds can often be reliable indicators of potential fertility problems—Table 9 lists some common weeds and the conditions they are associated with. You should also watch for signs of nutrient deficiencies, whether in weeds or cultivated crops. Table 6 lists some deficiency symptoms for all the important nutrients.

You may even gain a new sense of respect for your weeds once you have learned to carefully examine them. Although shunned as unwelcome invaders by most farmers, weeds are often less damaging to established crops than supposed. They can also provide many free soil-improvement services if allowed to pave the way for your desired crops. Perennial weeds in particular are frequently deep-rooted, helping break up hardpans, aerate the subsoil, and bring up mineral nutrients from areas too deep for crop roots. Or they may act as cover crops, protecting the soil from wind and rain until the space is needed. Some plants are such reliable mineral accumulators that prospectors can use

*Figure 11. Some common weeds. Sandy, droughty conditions are indicated by field horsetail, while red sorrel is a sign of wet, acid soil. Lambsquarters are abundant in fertile soil, especially where manure has been supplied.* (Acres USA Primer.)

them to indicate a possible source of deposits such as copper and selenium.

As you identify the weeds on your land, keep these considerations in mind:

- A few individual weeds don't mean much—look for consistent populations of the same species. The most reliable information comes from noticing several predominant species which all like similar kinds of soil conditions. For example, if you have a lot of plantain, chicory and coltsfoot, it is a clearer indication of heavy, poorly drained soil than if just one of those species was there.
- Perennial weeds, which regrow from the same roots each year, are better indicators of soil conditions than are annual weeds, which only survive one season.
- Many weeds can tolerate a wide range of conditions, but some are more specific in their requirements. Some indicate more than one factor—for instance, perennial sow thistles and docks both indicate wet areas, but the dock prefers more acid soils, while thistles like a higher pH.
- The growth characteristics of the weed can tell as much as its presence. Vigorously growing leguminous weeds, such as clovers, are a sign that soil nitrogen is low. Stunted,

*Table 9. Weeds & Soil Conditions*

**Bindweed, Field:** indicates hardpan or crusty surface with light sand texture.

**Bracken, Eastern:** indicates acid or low lime soil, low potassium and low phosphorus.

**Buttercup:** indicates tilled or cultivated soil.

**Buttercup, Creeping:** indicates wet, poorly drained clay soil.

**Campion, Bladder:** indicates neutral or alkaline pH.

**Chickweed:** indicates tilled or cultivated soil with high fertility or humus unless weeds are pale and stunted, then fertility is low.

**Chicory:** indicates soil with heavy clay texture, tilled or cultivated with high fertility or humus unless weeds are pale and stunted, then fertility is low.

**Cinquefoil, Silvery:** indicates dry soil often with thin topsoil, acidic or low lime.

**Colt's Foot:** indicates heavy clay soil, waterlogged or poorly drained, acidic or low lime.

**Daisy, Ox-eye:** indicates waterlogged or poorly drained soil that has been neglected or uncultivated, acidic or low lime with low fertility.

**Dandelion:** indicates heavy clay soil, tilled or cultivated, acid or low lime especially on lawns.

**Docks:** indicates waterlogged or poorly drained soil, acid or low lime.

**Grass, Quack:** indicates hardpan or crusty surface.

**Hawkweeds:** indicates acidic or low lime soil.

**Henbane, Black:** indicates neutral or alkaline pH.

**Horsetail, Field:** indicates sandy light soil, acidic or low lime.

**Joe-pye Weed:** indicates wet or waterlogged soil.

**Knapweeds:** indicates acidic or low lime soil with high potassium.

**Knotweed, Prostrate:** indicates tilled or cultivated soil, acidic or low lime.

**Lambsquarters:** indicates tilled or cultivated soil with high fertility or humus unless weeds are pale and stunted, then fertility is low.

**Lettuce, Prickly:** indicates tilled or cultivated soil.

**Meadowsweet, Broad-leaved:** indicates wet or waterlogged soil.

**Mosses:** indicates waterlogged or poorly drained soil, acidic or low lime.

**Mullein, Common:** indicates neglected uncultivated soil, acidic or low lime, low fertility.

**Mustards:** indicates hardpan or crusty surface, dry, often with thin topsoil.

**Mustard, White:** indicates neutral or alkaline pH

**Nettles:** indicates tilled or cultivated soil, acidic or low lime.

**Pigweed, Redroot:** indicates tilled or cultivated soil, high fertility or humus unless weeds are pale or stunted, then low fertility is indicated.

**Pineappleweed:** indicates hardpan or crusty surface.

**Plantains:** indicates heavy clay soil, waterlogged or poorly drained, tilled or cultivated acidic or low lime, especially on lawns.

**Shepherd's Purse:** indicates saline soil.

**Sorrel, Garden:** indicates waterlogged or poorly drained soil, acidic or low lime.

**Sorrel, Sheep:** indicates sandy light soil, acidic or low lime.

**Sow-thistle, Annual:** indicates heavy clay soil.

**Sow-thistle, Perennial:** indicates wet soil, neutral or alkaline pH.

**Stinkweed:** indicates hardpan or crusty surface, high lime.

**Thistle, Canada:** indicates heavy clay soil.

**Yarrow:** indicates low potassium.

---

yellowish weeds of other types may mean the same thing. The color of cornflowers will be blue on soils with a high pH and pink on acid soils.

- Each growing season, make note of any changes in the predominant weed species in your field records.

## Looking Beneath the Surface

The physical condition of your soil is its easiest aspect to evaluate. You don't need precise instruments to get a pretty good feel—literally—for its general tilth. Although your soil texture isn't likely to change much over time, the other tests are ones you may wish to repeat once a year or more—most are good indicators of how well your soil improvement program is working.

## Texture

Your soil's texture is a key factor in how well it can hold water and nutrients, and allow air to circulate. You can evaluate the texture by scooping up a handful of moist soil—adding a little water if it's very dry—and squeezing it in your fist. Now open your hand; if your sample falls apart instead of staying in a ball, you've got sand on your hands. If you have a ball of soil sitting on your palm, try to squeeze it upward with your thumb to form a ribbon. Make as long a ribbon as you can, and measure it. If you could make a ball but not a ribbon, classify your soil as loamy sand.

Now add water to your ribbon to make a soupy mud in your hand. Rub the forefinger of your other hand in it, and decide if it feels mostly smooth, mostly gritty, or equally smooth and gritty. Then match up your observations with the length of the ribbon you made on the chart below to find the texture which most describes your soil.

| | **Feels mostly gritty** | **Feels mostly smooth** | **Feels both gritty & smooth** |
|---|---|---|---|
| **Forms ribbon shorter than 1"** | Sandy Loam | Silty Loam | Loam |
| **Forms ribbon 1"-2"** | Sandy Clay Loam | Silty Clay Loam | Silty Clay |
| **Forms ribbon longer than 2"** | Sandy Clay | Silty Clay | Clay |

*[Adapted from* Methodologies for Screening Soil-Improving Legumes *by Sarrantonio]*

## Moisture

Once you know your soil's texture, you can more precisely determine how moist it is. Check moisture levels early in the growing season and any time you are concerned about the availability of water for your crops. Plants will wilt if soil moisture in their root zone drops below a range of 5% for sandy soils to 15% for clay soils. If soil moisture levels, as determined by the following method, are not above the 50% range within a 6 inch depth, many crops will suffer from lack of water. If your soil dries out to less than 50% moisture at this depth two days after a good rain, your soil is excessively drained.

First, look for obvious signs of surface crusting and cracking caused by dryness. See how far down you have to dig before the soil gets darker, indicating more moisture. Take handfuls of soil from different depths and squeeze them firmly in your hand. If your hand gets wet, your soil is saturated. Using the chart which follows, match the behavior of your handful of soil with the ap-

***Table 10***. *Guide for judging how much soil moisture is available for crops.*

| Available Soil Moisture Remaining | Feel or Appearance of Soil | | |
|---|---|---|---|
| | Light Texture | Medium Texture | Heavy Texture |
| 0 to 25 percent | Dry, loose, flows through fingers | Powdery dry, sometimes slightly crusted but easily broken down into powdery condition. | Hard, baked, cracked, sometimes has loose crumbs on surface. |
| 25 to 50 percent | Appears to be dry, will not form a ball.* | Somewhat crumbly but holds together under pressure. | Somewhat pliable, will ball under pressure.* |
| 50 to 75 percent | Tends to ball under pressure, but seldom holds together. | Forms a ball somewhat plastic, will sometimes slick slightly with pressure. | Forms a ball, ribbons out between thumb and forefinger. |
| 75 percent to field capacity (100 percent) | Forms weak ball, breaks easily, will not slick. | Forms a ball, is very pliable, slicks readily if relatively high in clay. | Easily ribbons out between fingers, has slick feeling. |
| At field capacity (100 percent) | Upon squeezing, no free water appears on soil, but wet outline of ball is left on hand. | Upon squeezing, no free water appears on soil, but wet outline of ball is left on hand. | Upon squeezing, no free water appears on soil, but wet outline of ball is left on hand. |
| Saturated | Water appears on ball and hand. | Water appears on ball and hand. | Water appears on ball and hand. |

*Ball is formed by squeezing a handful of soil very firmly.

[*Adapted from* Methodologies for Screening Soil-Improving Legumes *by Sarrantonio.*]

propriate texture and description to see how much water it contains.

## Drainage

A simple perc test will tell you if you have drainage problems. Dig a hole about a foot deep and 6 inches across. Now fill it with water and allow it to drain completely. As soon as it does, fill it again and see how long it takes to drain again. If it takes more than 8 hours, you should make drainage a top priority.

The rate at which water will infiltrate soil is a useful indicator of its porosity, and can be monitored with another free and easy on-farm test. The only tools required are a one-quart Mason jar, a tape measure, and a watch that can time seconds. In a representative part of the field, at a time when the soil is neither particularly wet nor particularly dry, empty the quart of water at soil level and start counting seconds. When the water has completely soaked into the ground, stop counting seconds and measure the diameter of the wet spot. Multiply the diameter by the number of seconds required for the water to soak in, and write the number in your field notebook or field records.

It doesn't matter whether measurements are in inches or metric, and farmers with clayey soils may prefer to use only a pint of water. What is important is that the volume of water is the same every time. If the soil is tested this way three or four times each summer and the results averaged, it will be possible to track subtle changes over the years. Declining numbers mean an improving ability to absorb water, reduce erosion and so on. Increasing numbers suggest that the system is deteriorating. Averages can be roughly compared from field to field (within the same soil type) but not from farm to farm.

## Structural Soundness

The size of the crumbs or aggregates that make up your soil, and how well they hold together, is a critical aspect of its tilth.

The bigger the crumbs, the bigger the pore spaces between them, and the better water and air can infiltrate. Soils with good structure can hold up to twice as much water as those with poor structure and the same texture. Plants grow better in soils with good structure, even when water isn't a problem.

First, simply pick up a handful of soil and see how it crumbles. Heavy soils with good structure will still crumble easily, and light soils will retain some shape instead of becoming powdery. Heavy soils with poor structure will resist crumbling or form large clods, while lighter soils will become dusty.

Next, dig a hole about a foot deep and examine the size of the crumbs at different depths. Look for large crumbs, about 1/32 to 3/8 inch in size. Check aggregate stability by placing a small handful of soil in a glass and filling it with water. Do the aggregates keep their shape, or does your sample collapse into a muddy blob? Good structure is indicated by a lot of large crumbs which hold together in a glass of water.

## Compaction

Compaction problems can be indicated by poor water infiltration, as shown by your perc test, as well as by poor structure. If you can't plunge a garden fork into your soil to its full depth without standing on it, assuming you haven't hit a rock or other obstacle, your soil is compacted. Another way to judge compaction is to shove a thin (1/4 inch) metal rod into the ground as far as it will go until it bends under pressure. Try this in various locations—the sooner it bends, the more compacted the soil. A more precise reading of compaction can be taken with a simple handheld tool called a penetrometer.

Scientific evaluation of soil compaction is measured by three criteria: **bulk density** tells you the weight of a given volume of soil, **pore space** indicates the proportion of a given volume of soil taken up by air and water, and **oxygen diffusion rate** measures how readily air passes through soil (see Table 4, page 21).

You can measure your soil's bulk density yourself, by following the instructions on page 66.

## Digging Deeper

To get a real worm's eye view of your soil, you might feel inspired to dig down a bit farther—two or three feet. If you have some project that entails digging a hole anyway, take a few minutes to make the following observations:

- Examine the layers of soil you see, and try to identify different horizons. The thickness and nature of each soil horizon tells a story about how the soil was formed, its drainage characteristics, and biological and chemical activity. The soil classification scheme described in Table 2, page 12 is based on characteristic features of different soil profiles. How thick is the topsoil? Are there sharply delineated layers below, with differences in color and/or texture? Mottled gray and rust brown streaks in lower horizons indicate leached, acid soils and possible drainage problems. Brown and red colors are from oxidized iron, a result of acidity. Absence of definite horizons may indicate very deep topsoil, a muck or organic soil, or a young soil that has not been subject to a long weathering process. Layers of different colored clays are commonly found in tropical soils.
- How deeply do plant roots penetrate? How far down do you find worm tunnels and castings, or other visible creatures?
- Note the depth of any compacted layer or hardpan, where roots end abruptly.
- Note the depth of any area of bedrock, gravel, or other kind of stone.

A key component of soil vitality is the movement of water, soil life, organic matter, and nutrients between the topsoil and the subsoil. The nature of the topsoil-subsoil boundary is a wonderful indirect test that summarizes multiple factors.

If the topsoil is largely uncoupled from the subsoil, the boundary between the two will be distinctly visible, and relatively even. Very little organic matter or microbial activity crosses such a barrier, which is usually caused by cultivation methods. In a vital

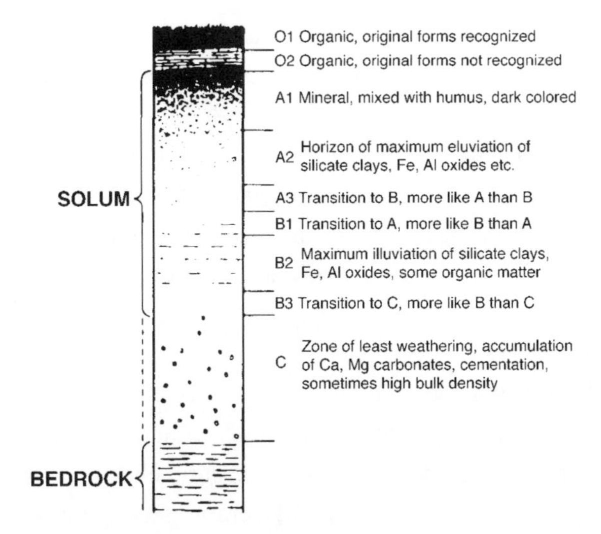


***Figure 12.*** *Soil Horizons. "O" layers are topsoil; "A", "B", and "C" layers are subsoil. It is sometimes easy to see clear differentiation between horizons. In Spodosols (characteristic of the Northeast), there are distinct color changes where minerals leached from the "A2" horizon are precipitated in the "B2" horizon. (Reprinted with permission of MacMillan Publishing Company from The* Nature and Properties of Soils, *8th Ed., by Nyle C. Brady. Copyright © 1974 by MacMillan Publishing Company.)*

soil, by comparison, the topsoil-subsoil boundary is quite uneven and fairly gradual. Roots penetrate to greater depth and when they decay they leave darker channels that couple the topsoil to the subsoil.

Once you've caught the soil exploration bug, you may want to pursue a few more sophisticated observations. The following measurements require simple tools, and in some cases involve waiting a while to see the results.

## Erosion Watch

If you are concerned about soil loss on a slope, there are a couple of methods which can confirm that erosion is taking place. The first involves driving a hollow cylinder into the soil—a 6″ piece of plastic sewer pipe works well—so that the soil level inside the cylinder is the same as that outside. Place it on the upper end of the plot in question. Cover the cylinder and check it periodically. If the soil level outside gets lower than that inside, you will know that soil is being lost.

The second method involves sinking a pie pan into the soil anywhere on the slope. Angle it in line with the slope, and make the lip level with the soil surface. Check after a heavy rain to see if you have caught some runaway soil in your pan.

## Bulk Density

Although considered a technical observation, periodic monitoring of bulk density is among the most conclusive indicators of changes in soil tilth. It isn't hard to measure: choose a spot that is relatively undisturbed, and start by scraping away the surface to level it a bit. Then dig a hole about 6 inches deep and 4 inches in diameter, saving all the soil you take out in a bag, and smooth out the bottom of the hole.

Now stuff a plastic bag into the hole and fill it up with a measured amount of water—use a kitchen measuring cup to dip it out of a bucket. Be sure the water is as close as possible to the soil surface, and record the total volume of water it took to fill the hole (use metric units to be able to compare the figures with scientific data). This figure gives you the volume of soil you removed.

Next you have to dry the soil by spreading it on a cookie sheet in the oven with the setting on "warm" for a day or so. Then weigh it. If you are using metric figures, you need a scale that reads in grams. You can now calculate the bulk density (B.D.) by dividing the weight of the soil by the volume. A good range is 1.0-1.8 grams per cubic centimeter. More than 2 grams per cubic centimeter is a sign of compaction.

## Field Capacity

Field capacity is a measure of how much water your soil can hold when it is saturated—increasing field capacity is another sign of increased humus content. Do this a day or two after a soaking rain, or you can flood a small area first; wait until the water has soaked down into the lower levels. Now dig up a few samples of soil from areas two to four inches down—an ounce or two is enough.

Weigh the sample you collected, then dry it the same way outlined for bulk density, above. Now weigh the dry soil—make sure to subtract the weight of the container it is in each time.
The figure for field capacity is expressed as a percentage, calculated as:

$$\frac{\text{wet weight of soil - dry weight of soil}}{\text{dry weight of soil}} \text{ X } 100\% = \text{Field Capacity}$$

For comparison, well granulated silt loam has a field capacity of about 15%.

## Evaluating Biological Health: Looking at Life Forms

Any time you work in the soil, you should be noticing how biologically active it is. Try to identify the kinds of macroorganisms you see crawling, skittering and squirming around as you transplant or hand weed. An inexpensive hand lens can open up a whole universe of creatures you never knew existed.

### Organic Content

There are also more pointed observations you can make to evaluate a potential garden site or to track improvements in your soil's biological activity. It's a good idea to have organic matter content measured when you have your soil tested (see page 74), but you can visually evaluate it yourself. In general, darker browns mean higher humus content. Look for white threads of fungal mycelia and bits of undecomposed organic matter. If your soil is biologically active, you will not be able to recognize crop residues from the previous season within a few weeks of spring planting. If you still find old cornstalks and broccoli stems in June, or a layer of sod from last fall's tillage, you need to work on stimulating soil organisms.

## Counting on Worms

An earthworm census is a valuable barometer of soil biological activity and overall system health. When you count your worms, be sure to avoid examining places which would distort the results, such as near a compost pile or under mulch. Also, do this during a cool time of day, to avoid exposing worms to harmful conditions. Record your earthworm estimates several times per season, and use the season average to gauge changes in soil health from year to year. Use any of these methods for estimating the earthworm population of your soil:

- Measure out a foot-square plot and dig down 6 inches, placing everything you dig up in a bucket or pan. You should find at least ten earthworms in that much good healthy soil.
- Take a "spade split" of soil, about two inches thick and eight inches deep. Each earthworm in the spade split represents about 100,000 worms per acre. Take several spade splits per field, and record the average estimated earthworm population.
- Simply count the number of earthworm holes, with their little piles of castings next to them, observable in the same well-delineated area over the course of a season, and compare that number from year to year.

## Rooting for Healthy Soil

The condition of plant roots is one of the ultimate indicators of soil health. Observe the root systems of weeds as you pull them up. You can also dig up a plant just to examine its roots, being careful to cut off as little as possible of its root mass. Choose the healthiest looking specimen in the vicinity, and try to answer the following questions:

- Do the roots seem vigorous and well branched? You want to see roots penetrating as much of the soil area as possible—track the approximate volume and depth of root mass in a given area as an indicator of changes in biological health.
- Are there lots of fine root hairs? Finding only a few fine feeder roots might indicate that too little air is available in the root zone.
- Do roots grow freely in every direction, or do they seem to grow sideways at a certain depth? Roots which suddenly turn sideways are a sure indication of a hardpan.

The root systems of legumes, such as clover, vetch and garden beans, should contain a good sprinkling of nodules. This is where the nitrogen-fixing *Rhizobia* bacteria live. If you slice open some of these nodules, they should be noticeably pink or red in color. A green or black color indicates a lack of bacterial activity. (This may be a normal phase if the plant is entering dormancy.) The size and number of nodules is directly related to how much nitrogen the plant can fix. Different species of legumes normally differ as to rates of nodulation—alfalfa and clover are generally more heavily nodulated than beans or peas. Using the right bacterial inoculants on your legume crops will increase nodulation.

Regular observation and recording of as much information as possible about your soil's physical and biological condition, in addition to your records of crop health and yields, will give you a solid basis for evaluating the progress of your fertility improvement program. This, combined with chemical analysis of soil nutrient content, will help you achieve optimum soil health and crop productivity.

## Soil Nutrient Testing

Much of the soil testing in ecological farming should be focused on monitoring the vitality of the whole system and its components. Soil nutrient testing is simply one component of the farm management and decision-making process. In particular, it is a means of *monitoring* conditions and progress towards a goal.

*Root system of a healthy legume, showing nodules containing symbiotic, nitrogen-fixing rhizobia bacteria.* (USDA Soil Conservation Service)

The accuracy of the soil test and resultant recommendations depend on many factors including sampling procedure, analytical process and laboratory "philosophy" in interpreting the results. If these factors are fully understood, the soil test can be a useful, if limited, guide to ecological soil management.

A farmer or gardener has three soil-testing choices: a home test kit, a government or university test, or a private laboratory. Those with a natural scientific curiosity may be well served by the purchase of a soil test kit. Useful kits range in price from $45 to $150. These kits will measure nitrogen, phosphorus, potassium, pH, and often organic matter, calcium and magnesium. The chemical reagents and simplified procedure normally allow "ballpark" accuracy. Cheaper kits may lead to misleading results.

University or government tests tend to be limited in scope, but are often backed by "field response" data (see Figure 13, next page). This research correlates yield differences with fertilizer applications for specific soil types and test results. The resulting data are used as a basis for fertilizer recommendations. Some private labs, such as A & L Agricultural Laboratories, offer more elaborate tests, such as base saturation ratios, cation exchange capacity, and quantified results (as compared to "high, medium, low"). These soil audits are more expensive than university tests. (Figure 14, page 73, is an example).

## Testing Procedure

Soil sampling procedures are often the most important part of soil testing. Ironically, they are also the area of greatest weakness. Sampling information is available from any private or public soil testing lab. Take all recommendations about getting a representative sample seriously; a surface sample from only a small area of a field will be misleading.

Even a carefully taken sample is not an accurate picture of actual field conditions. Chemical reagents do not interact with soil the same way that plants do. Your sample is only to plow

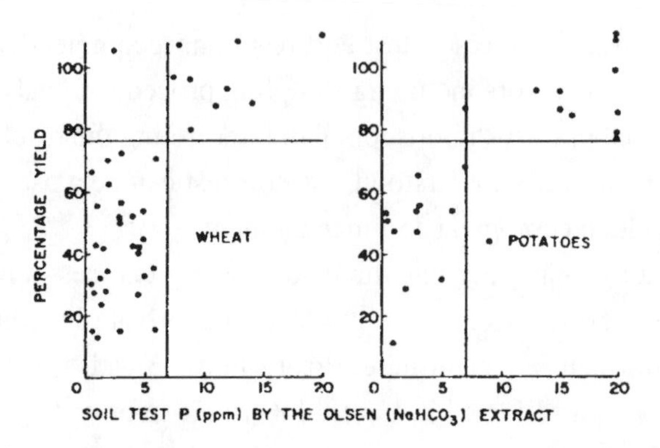


*Figure 13. Field Response Data Illustrated. Each dot represents a field test plot. The distribution of responses shows a greater response to higher soil phosphorus by potatoes than by wheat.*

depth, whereas deeper levels of soil can provide significant nutrients. Moreover, soil chemistry can be quite different at different times of year and moisture conditions, and may even vary from day to day. For example, researchers at the University of Vermont have found significant differences between the reactions of soils that are dried prior to testing and those that are tested moist. This perspective should be considered whenever using soil testing to make management decisions.

The calculations that appear on laboratory report forms look impressively precise, but the degree to which those numbers represent conditions in your soil can range from "pretty good" to "anybody's guess." The analytical procedures used in labs vary enormously. Some chemical reagents and procedures are appropriate for certain soil conditions but not for others. In addition, in order to keep costs low, some labs sacrifice analytical precision by using less expensive equipment or less labor-intensive methods. Various errors, both human and computer, can also occur. There is also a lack of standardization or inter-lab agreement about soil testing procedures in general.

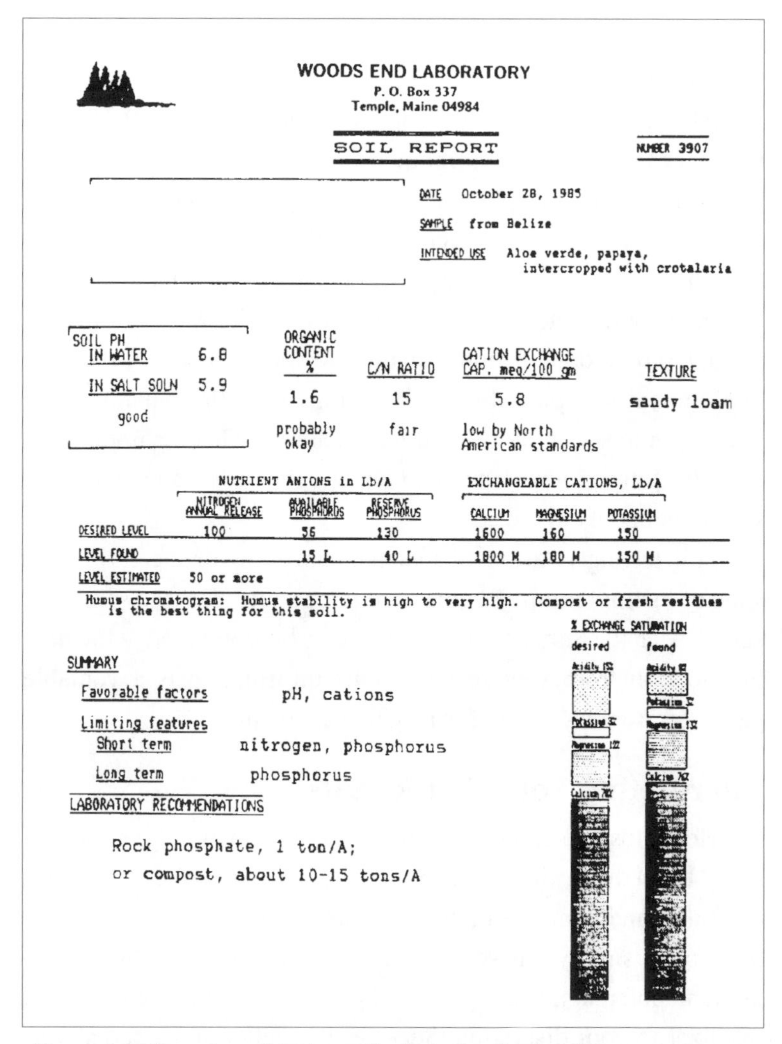

WOODS END LABORATORY
P. O. Box 337
Temple, Maine 04984

SOIL REPORT

NUMBER 3907

DATE October 28, 1985

SAMPLE from Belize

INTENDED USE Aloe verde, papaya, intercropped with crotalaria

| SOIL PH | | ORGANIC CONTENT % | C/N RATIO | CATION EXCHANGE CAP. meq/100 gm | TEXTURE |
|---|---|---|---|---|---|
| IN WATER | 6.8 | | | | |
| IN SALT SOLN | 5.9 | 1.6 | 15 | 5.8 | sandy loam |
| good | | probably okay | fair | low by North American standards | |

| | NUTRIENT ANIONS in Lb/A | | | EXCHANGEABLE CATIONS, Lb/A | | |
|---|---|---|---|---|---|---|
| | NITROGEN ANNUAL RELEASE | AVAILABLE PHOSPHORUS | RESERVE PHOSPHORUS | CALCIUM | MAGNESIUM | POTASSIUM |
| DESIRED LEVEL | 100 | 56 | 130 | 1600 | 160 | 150 |
| LEVEL FOUND | | 15 L | 40 L | 1800 M | 180 M | 150 M |
| LEVEL ESTIMATED | 50 or more | | | | | |

Humus chromatogram: Humus stability is high to very high. Compost or fresh residues is the best thing for this soil.

SUMMARY

Favorable factors pH, cations

Limiting features

Short term nitrogen, phosphorus

Long term phosphorus

LABORATORY RECOMMENDATIONS

Rock phosphate, 1 ton/A;

or compost, about 10-15 tons/A



*Figure 14. Sample soil test results for a tropical soil.*

## Fertilizer Recommendations

There are no universal standards for optimum mineral levels for all soil and crop situations. A representative sample, analyzed by appropriate testing methods, yielding accurate numbers, can still bring different fertilizer recommendations from different labs. Some labs accept certain standards for sufficient level of available nutrients (S.L.A.N.). Others recommend higher levels for "maintenance" of minerals, while many private laboratories use the Albrecht formula of base cation saturation ratio (B.C.S.R.). Often these methods are combined for different nutrients or soils. Regardless of philosophy, labs base their interpretations on their experience with previous recommendations.

The value of soil tests will depend on continued research which considers the effect of different fertilizer rates not just on yield, but on crop nutritional quality. In some schools of thought, less emphasis is placed on soil testing. The Bio-dynamic view stresses humus quality over mineralization. Others maintain that plant feeding is more dependent on energy flows than the level of soluble minerals in the soil.

Soil testing is a useful management tool for the farmer, in spite of its limitations. Regular soil test results from the same lab, sampled at the same time of year, can be compared with field information from year to year. This information is a valuable gauge of progress in soil fertility management.

## Interpreting Your Test Results

How closely should you follow the information given in soil tests? Read the fertilizer recommendations and explanatory notes carefully, and compare them to nutrient levels discovered. Some lab reports simply tell you how much nitrogen, phosphorus, potassium, and magnesium you need per acre, leaving the choice of fertilizer to your discretion. Services that tell you the actual versus desired level of nutrients allow you to check fertilizer recommendations for arithmetical errors. If you don't understand why any recommendation is made, contact the testing service. Recommendations are often based on your intentions for the field. If you change your plans, nutrient requirements may change as well. (Further details about fertilizer use and availability are given in Chapter 4.)

**Cation Exchange Capacity & pH:** Both of these figures indicate how effectively your soil is able to store nutrients and release them to plants. Both are fairly accurate, but pH values can vary by about half a unit, depending on whether the test is done in water or salt solution. Recommendations should be aimed at attaining a pH of about 6.5 for general purposes.

Other than adding lots of organic matter, there is little that can be done to significantly alter the cation exchange capacity. But the CEC does have a bearing on how much fertilizer you'll need. If the soil's CEC is derived mainly from its organic matter rather than its clay content, raising a low pH can improve the CEC as well. For every increase in the CEC, more cations (generally calcium in lime) will be required to raise the pH the same amount. If you have a high cation exchange capacity but depleted mineral reserves, more fertilizer is needed to satisfy crops, since the soil reserves will keep some of the cations unavailable to crop roots.

**Major Cation Nutrients:** Testing procedures for calcium, magnesium and potassium are quite accurate, and the results are reliable indicators of soil needs. Private labs subscribing to the Albrecht formula will indicate relationships between these elements and acidity (percent base saturation). This will tell you whether general nutrient levels need to be increased across the board, or whether one element in particular is lower than the others.

The best practice is first to bring all nutrients into balance, and then increase their overall levels—up to the optimum desired—while keeping the pH around 6.5. You can do this by looking up the analysis of the fertilizer you plan to use and multiplying the percentage of a given component by the number of pounds you will apply per acre. Then compare the result with the difference between desired and actual levels shown for that nutrient. For example, if you are using Sul-Po-Mag™ (which contains about 22% $K_2$, 18% MgO, and 22% $SO_4$) and want to add about one hundred pounds of potash per acre, you should use approximately five hundred pounds of Sul-Po-Mag™ per acre. However, you

## Calculating Cation Fertilizer Needs

Laboratory recommendations for cation fertilizers are given in pounds per acre. The chemistry upon which those recommendations are based depends on the **valence** of the element in question, not its weight. Valence refers to the charge of an ion and the number of electrons it accepts or gives up when forming a neutral bond.

Of the major cations, calcium ($Ca^{2+}$) and magnesium ($Mg^{2+}$) occupy the same number of cation exchange sites, though calcium weighs a little more than magnesium (by a factor of 1.2). Therefore, you need fewer pounds of magnesium than of calcium to neutralize acid conditions. In other words, when considering the need for a good soil cation balance, you should expect to find about ten times as much calcium as magnesium *by weight* in test results. This corresponds to a base saturation ratio of six or seven to one.

Potassium ($K^{+}$) has only half the chemical charge of calcium or magnesium, and so requires twice as many ions to fill the same number of cation exchange sites. One ion of potassium, moreover, weighs a good deal more than one of magnesium. Therefore, although magnesium occupies three to four times the number of soil cation exchange sites as potassium, it is required in less than twice the amount of potassium by weight.

would also be adding about fifty pounds of magnesium (in the form of MgO) per acre, so be sure that this won't send your potassium:magnesium balance out of whack.

**Organic Matter & Nitrogen:** The quality of organic matter—that is, its balance between fresh, decomposing forms and stable humus—is more indicative of soil health than its quantity. Monitoring organic matter quality is more an art, developed through observation and experience, than a science. Measuring organic matter quantity can nevertheless be of value for comparing changes over time in the same field, especially when combined with observations of soil structure, porosity and water holding capacity.

Several cautions are in order. Readings of organic matter quantity may vary considerably, depending on which test is used. One common test is combustion, in which a dry soil sample is weighed, burnt, and weighed again. The difference is assumed to be organic matter. There are also several chemical digestion methods, and some tests that measure "humus." Make sure that the testing methods are the same before trying to compare results. Similarly, organic matter is usually highly variable within a field, making consistent sampling procedures very important.

Nitrogen is considered the least reliable of chemical tests because of fluctuations in soil nitrogen levels and availability. Some labs simply estimate nitrogen at about 5% of organic matter, which can vary as noted above. For these reasons, many labs do not test for nitrogen, although a test recently developed by the University of Vermont has proven to be more accurate.

Nitrogen recommendations are generally based on crop need, rather than existing reserves in the soil. You must take into account field history factors such as contributions from leguminous crops and green manures, varying nitrogen uptake of preceding crops, and previous applications of manures and other organic matter.

Annual nitrogen release (mineralization) can be estimated by considering humus stability, along with total organic matter content and average annual temperature for your region (which influences the rate of microbial activity). Improving the level of organic matter will improve soil productivity. Proper management of organic matter is a never-ending necessity—no matter how positive your test results are.

**Phosphate:** Phosphate is another complex nutrient; it is uncertain how accurately testing procedures reflect growing conditions and the variety of factors influencing phosphate availability. Again, the numbers appearing in soil test results shouldn't be too heavily relied upon for calculating fertilizer needs. Since ecological practices stress building reserves that are made available *gradually* through organic matter decomposition, you are quite safe in applying as much rock or colloidal phosphate as you can afford if the phosphate level tests low.

If organic matter content is low, using fish emulsion to "water in" transplants, or for foliar feeding (see Chapter 4) will boost available phosphate. If you choose acidulated phosphate fertilizers, how much to use depends on the immediate crop demands rather than soil-building considerations. Of course, the pH and organic matter content of your soil strongly influence the availability of reserve or applied phosphate to plants, and are therefore just as important as measured soil phosphate levels in ensuring adequate phosphate levels for crops.

**Minor & Trace Elements:** In general, as the level of an element declines in a sample, testing procedures become more expensive and less reliable. Plant tissue tests are a better means of evaluating micronutrient nutrition. Tests are readily available for sulfur, boron, copper, manganese, zinc, and iron; of these, the most extensively researched is for boron. Such tests are most valuable when you plan to grow a crop with a high demand for some specific minor element (such as boron for alfalfa or brassicae), or have cause to suspect a deficiency. In the case of a deficiency, use micronutrient supplements with extreme caution to avoid undermining crucial balances.

**Soil "Hazards":** Some soil tests also indicate the level of exchangeable aluminum ions, often excessive in the Northeast. Under conditions of low pH, aluminum becomes more available—a problem because of its toxicity to plants and the fact that it interferes with the availability of nutrients like phosphate. Raising the pH by liming eliminates the hazard. It is also possible to have your soil tested for toxic heavy metals, such as lead, cadmium, selenium, and mercury. Check with your nearest university soil lab to find out how.

If the farm has a recent history of using agricultural chemicals, then testing soil, plants and water for residues of these chemicals may be an important means of monitoring how well the farm system is beginning to "clean itself out." If the farm has been in hay for the last twenty years, residue testing is probably less important, but it should be remembered that some of the early pest control products (such as lead arsenate and DDT) leave significant residues for decades. Because both residue sampling and testing are complicated (and therefore expensive) procedures, competent professional assistance is usually advisable.

## Tissue Analysis & Other Tests

By complementing a soil test with a tissue (usually leaf or petiole) analysis, a more complete picture of soil fertility emerges. Parameters for nutrient levels in many plants, at different growth stages, have been established. Plant root systems draw nutrients from a large area of topsoil and subsoil. Plant tissue analysis reveals a lot about levels of soil nutrients, as well as that plant's ability to obtain them. It is important that the tissue sample is taken from the correct areas of the plant and at a growth stage for which parameters are established.

Crops are often tested after harvest for nutrient and energy content. This information is generally used for balancing livestock feed rations. However, it also indicates how available soil nutrients are to crops—another indicator of soil fertility.

Some agricultural consultants also test the plant's nutrient status with a **refractometer**, a hand-held instrument that measures the sugar content of cell sap. Sugar content is closely related to plant mineral content. After placing a few drops of sap on a glass plate, a visual display tells the percent of sugar content according the **brix scale**. An experienced consultant can estimate plant and soil health by comparing the results to established parameters.

Soil tests, tissue analysis, indicator weeds, test strips, and refractometer checks are among the tools available to give you feedback about your own observations of soil health. Many other innovative testing techniques are also available for the farmer interested in exploring the frontiers of farm system monitoring. These are frequently discussed in publications such as *Acres USA*, and may assist the monitoring and evaluation process considerably.

The important thing to remember about soil testing is that it involves far more than laboratory analysis. Laboratory work is a useful supplement to the more important on-farm testing and observation, but lab results alone are of limited utility for a management system that attempts to do more than achieve simple target yields.

When coupled with sound and holistic farm management, thoughtful monitoring to learn how clean, vital and diverse the farm system is, will contribute significantly to making the farm healthy and prosperous.

# 4

# Soil Management Practices

## Managing Physical Factors: Drainage, Irrigation, Tillage, & Cultivation

In planning the most appropriate soil management practices the first factor to consider is soil moisture and soil aeration. The most sophisticated fertilization program will give scant return if the soil does not have a proper moisture/air balance. This balance will not be present if the land is poorly drained or does not have adequate moisture. Organic farming relies on conventional drainage practices if these are ecologically sound, complying with marshland conservation measures. Organic management in arid areas can even drought-proof soil to some extent. Tillage practices will vary depending on the soil and climate but the principle is to preserve organic matter in the top horizon. In perennial situations permacultural practices would largely replace annual tillage and most cultivation. Organic farmers have been the leaders in developing innovative mechanical cultivation systems for weed control, but that only serves to supplement control based on a well-tuned rotation.

### Drainage

Farming in areas where the water table is high may require subterranean tiling with perforated pipe or a gridwork of ditches. Heavy clay soils and those with subsoil hardpan layers may require special measures, such as subsoiling with a bulldozer, to allow better water and root penetration. Many soils that drain poorly simply require ecological management to improve percolation and break up plow pans and compaction, but structural problems may require drainage. Consult your Extension or Soil Conservation Service for the most appropriate local services.

## Irrigation

It is not within the scope of this book to discuss the ecological ethics or practical implementation of irrigation in organic farming systems. Suffice it to say that if irrigation is appropriate on any farm it should be managed so that water use is optimized. In most practice that would mean that flood or sprinkler irrigation would not be as acceptable as drip irrigation. The principle of ecological management is to improve the water retention capability of soils and design the farming system to optimize the water cycle. This is accomplished in arid areas by following guidance offered in systems such as Permaculture, Holistic Resource Management or Keyline Management, not in re-routing rivers to desert areas.

## Tillage

The function of tillage is to incorporate organic matter and mineral fertilizers into the soil, aerate the soil, improve water permeability, control weeds, and prepare a seed bed. Since all tillage destroys organic matter and causes compaction to some degree, tillage systems should be designed for minimal use and damage to soil structure.

A tillage system should work crop residues and other organic matter into the topsoil where, in the presence of bacteria and other life forms, it can be digested. Some sort of vegetative cover or litter should be left on the soil during the northern winter, southern dry season or tropical rainy season. A properly managed tillage system creates a seed bed where carbon dioxide, oxygen, nitrogen, and water circulate through the soil, fostering the biological processes that create soil fertility and plant growth.

Seed beds are often created in a way that damages soil structure resulting in erosion, compaction, and organic matter oxidation. For example, one common conventional method is fall moldboard plowing, spring discing, and a final harrowing. The plow can create a hardpan with annual use, bury organic matter and living topsoil in an anaerobic zone, allow bare subsoil to erode by wind and water, and obstruct the capillary action of water. Spring discing breaks down the clumps, but it also causes compaction especially in wet conditions. The field may look good after the final harrowing, but not without great expense to soil tilth.

Properly designed tillage tools avoid these problems. Although the moldboard plow is useful to turn a heavy sod, it is inappropriate for most farms. New types of plows, such as the English paraplow, Australian Keyline plow and the German bioplow, can accomplish plowing tasks without the detrimental effect of conventional plowing. Many organic farmers base their system on the chisel plow for deep tillage in the fall and off-set discs, such as Miller, for seed-bed creation in the spring. Rotovators are used by a number of farmers, especially market gardeners. While they do mix organic matter into the top layer, they can create a hardpan and leave the soil too fluffy. They are best employed on a sandy soil with a subsoiler or heavy chisel used annually to break the hardpan.

In response to tillage problems, and in recognition of the need to nurture a living soil, a number of combination tools such as the Lely Roterra, Glencoe Soil Saver, Landoll Soil Master, and Weichel Terravator have been developed. These tools have their proponents and have grown in popularity since the early 80s.

The other development in conservation tillage tools has been the Ridge Till and Zone Till systems. These tools, developed and promoted by companies like Buffalo Farm Equipment, create field ridges for overwintering and spring planting. Ridging seems to decrease erosion, create a dynamic seed bed, and allow for easy weed control. No-till systems usually rely on herbicides and hence are unsuitable for organic farming.

Ecological tillage is as much a question of timing as it is one of proper equipment. Only experience with your own soil and climatic conditions, combined with knowledge of equipment capabilities, will lead to an appropriate tillage system for your farm.

## Cultivation

The main reason for cultivation is weed control. Cultivation also breaks soil crusts and stimulates biological activity through aeration. However at the same time it oxidizes organic matter and damages soil structure.

Weeds are a major challenge for organic farmers. Weeds are controlled but never eliminated under ecological management systems. Knowing the life cycle of problem weed species can enable you to decide on the best method of control. A well tuned rotation is the basis of ecological weed control. Aspects of rotations which serve to control weeds include:

- Mixing sod and row crops in rotation; alternating control methods for warm and cool season weeds; matching crop needs to nutrient levels, especially nitrogen, so that excesses do not spur weed growth.
- Green manure smother crops like buckwheat, and allelopathic crops like rye, which excrete exudates that suppress competition.
- " Living mulches" or legume intercrops that smother weeds as well as fixing nitrogen and adding organic matter. Sometimes weeds themselves can function as a mulch if they do not set seed or spread.
- In perennial settings timely mowing can control weeds if shading or competition do not.
- Animal grazing, especially when optimized with moveable fencing. (See pages 113–122 for more on rotations.)

Biological controls, such as weed-eating insects and weed diseases are becoming more popular, as chemical companies diversify into bio-rationals and introduce ecological products. Plastic mulch of varying colors and degradability are also allowed to or-

*The Glencoe Soil Saver—combination tillage tools reduce the number of passes across the field.*

ganic farmers with restrictions that include proper post-season clean-up.

These practices do not usually eliminate the need for some mechanical cultivation to control weeds—many farmers feel that there is no substitute for fast moving steel. Innovative farmers and manufacturers keep developing increasingly effective equipment. One of the most popular implements for organic farmers is the rotary hoe. This is a tool bar mounted with circular steel tipped wheels on individual springs. It is pulled at high speed over crops when they are two to three inches high and kicks out weed seedlings without harming established plants. Farmers have continued to modify this tool to their needs, some welding on heavier tips and trash guards to convert it to heavier tilling functions, while others have added a second tool bar and holes to double its action and cut down on field trips.

Harrows have also been modified and put to uses other than finishing the field for planting. For example, light harrows are used to rake a field after planting (termed "blind harrowing") to disrupt weed growth. Row cultivators have also undergone a radical change in design. The Bezzerides brothers developed attachments for cultivators that replaced discs and sweeps with spyder, thinner, spinner, and weeder tools which work the soil lighter and closer to the crop. Other innovative implements include the Lilliston rolling cultivator, Rau-Kombi Vibra-shank, and the Lely weeder. Ridge tilling also has its own style of cultivating equipment.

There are other approaches available when fast steel isn't appropriate. The Bärtschi company has developed an effective brush weeder by which vertically mounted polypropylene brush wheels scrub between crop rows. This tool works well in wet weather and doesn't bring up new weed seeds by penetrating the soil deeply. Thermal weed control, always popular in Germany has become better known in North America. These liquid petroleum powered flamers are ecologically appropriate in many situations.

In citrus groves where plastic irrigation piping prevents mechanical and flame weeding, a hot steam applicator is proving successful in its trial season. Electric shock machines, once popular in western sugar beet operations, are very effective on large, laser-leveled fields. There are also weeders that use mounted rubber tires to pull out established weeds. Sod crops can also be helped by mowing, harrowing, and aerating operations.

All organic weed control methods, however, are based on timing and a healthy soil. The entire tillage system has to be timed to give the crop a competitive advantage over weeds. Some farmers eliminate fall tillage and/or delay spring tillage to plant when temperature and moisture are ideal for the quick germination and growth of crops. A healthy soil is not as prone to serious weed problems and provides a granular seedbed that insures the rapid emergence and vigorous early growth of the crop.

## Organic Matter & Humus Management

### Building & Maintaining Humus

The most important task of any ecological soil management program is creating and maintaining humus. This is not to say that other concerns, such as mineral nutrient levels, should be neglected. However, while it is possible to promote healthy soil by carefully managing organic matter and humus with little regard for balancing minerals, the reverse is rarely true.

The conventional and ecological approaches both consider organic matter important for soil fertility, although they assign a different priority to it. But organic matter in itself is not enough. It must be continually undergoing biological transformation into stable humus. This process is entirely dependent on a healthy microbial population.

The optimal form, quantities, frequency, and timing of organic matter supplementation will be determined by environmental, management and economic factors.

*A rotary hoe.*

**Environment:** In general, warmer climates coincide with lower soil organic matter levels. Where lack of moisture isn't a limiting factor, higher temperatures cause organic matter to decompose faster than the corresponding increase in vegetative growth. Tropical soils contain little organic matter because it mineralizes as fast as it drops to the ground. In hot, dry (arid) climates, decomposition is very slow, but so is vegetative growth that would provide fresh organic matter.

Drainage and soil texture are also important. Waterlogged soil favors anaerobic decomposers over the faster aerobes. When moisture is sufficient, sandy, well-aerated soils promote faster microbial decomposition than do heavier clays, which also tend to hold onto their organic fractions more securely. Soils under grasslands in temperate climates will accumulate more organic matter than forest-covered soils of similar texture and climate zones.

**Management:** Within the context of commercial agriculture, essential elements for protecting and increasing soil organic matter include sod or other permanent ground cover in the temperate zones, where the soil has the benefit of a period of conservation during the winter. In tropical climates without that rest period, a system of mixed annuals and perennials should be employed. The more a soil is worked, the faster its organic matter will be exhausted—one reason for the promotion of minimum tillage techniques. Simply preventing erosion will eliminate one of the major causes of organic matter depletion.

Soil that suffers from compaction and poor drainage will lack the microbial populations needed to "process" organic matter; any organic matter that is added to this soil may just sit there, unavailable to crops, and with no stable humus to keep soluble nutrients from leaching away. Carefully managed cultivation techniques can keep organic matter plentiful and productive—rather than exhausted or inactive.

*Discing in barley stubble, leaving a residue of 2,200 lbs. per acre. (USDA Conservation Service.)*

*Table 11. General crop nitrogen requirements.*

| Crop | Pounds of nitrogen per acre per year |
|---|---|
| Grass (2-3 times as a top dressing) | 100–150 (maximum) |
| Small grains | 20–40 |
| Potatoes | 120–160 |
| Leafy vegetables | 120 |
| Root crops | 80 |
| General home vegetables | 100 |

***Note:*** *These are the guidelines used by the University of Vermont in making nitrogen fertilizer recommendations, before taking into account soil test results or past cropping history.*

## Organic Matter & Nitrogen Availability

It is often helpful to plan organic matter management by linking it to crop nitrogen needs. (See Table 11 for general crop nitrogen requirements.) Nitrogen is made available gradually as organic matter mineralizes, at a rate determined by the level of soil biological activity and the type of organic matter present. The addition of green manures, legumes in rotation, animal manures, and compost will supplement existing soil nitrogen levels. Carbonaceous organic matter, such as straw, sawdust, and cereal stubble will temporarily deplete soil nitrogen.

It is impossible to calculate precisely how much available nitrogen is being contributed by organic matter, but information about the amount and nature of soil organic matter, as well as the approximate rate of decomposition you can expect, can help you estimate how much of your crop's nitrogen needs can be met through organic matter. Figure 15, next page, shows the relationship between available nitrogen, soil organic matter levels and annual rate of decomposition. Note that as soil organic matter levels increase, the rate of mineralization (k) tends to decrease. Highest levels of available nitrogen occur in soils with about 3-4% organic matter, which decomposes at a rate of 3-4% per year.

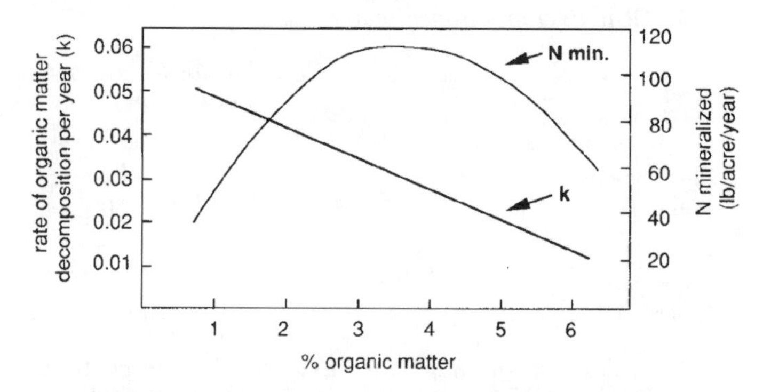


*Figure 15. Nitrogen availability and its relation to percent organic matter and decomposition rates. (Redrawn from F.R. Magdoff,* Communications in Soil Science, *1991, p. 1512. Courtesy of Marcel Dekker, Inc., New York.)*

The following example shows how you might estimate nitrogen availability from organic matter:

Assume the weight of soil in the top 6 inches of an acre equals 2,000,000 pounds. Your soil test indicates an organic matter content of 5%, so the weight of organic matter per acre is 100,000 pounds. Nitrogen comprises about 5% of organic matter, so the total nitrogen content of each acre will be 5,000 pounds. As indicated in the figure, you can reasonably expect a decomposition rate of about 2% a year, resulting in about 100 pounds per acre of available nitrogen from organic matter—enough for the needs of most vegetable crops.

This example also tells you how much organic matter you will need to add in order to replace what's lost through mineralization: 2% of 100,000 pounds, or about one ton. Some of this may come from crop residues, but additional compost, manure, or green manure in rotation will be required. You should note that this example does not account for differences in climate, moisture levels, or soil texture, all of which influence decomposition rates.

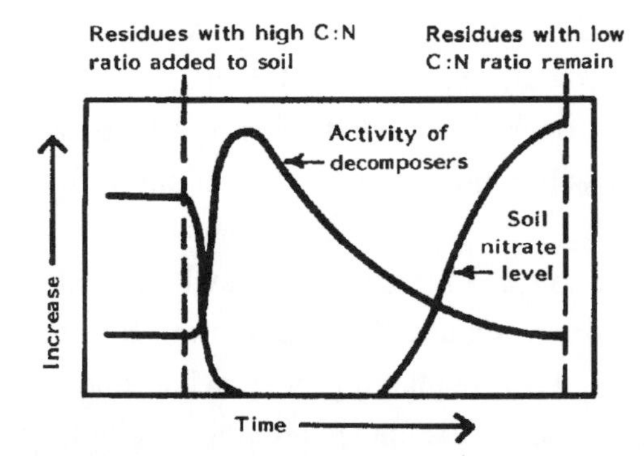


*Figure 16. The Effect of Carbonaceous Materials on Soil Nitrogen. When highly carbonaceous (high C:N ratio) materials are added to the soil, decay organisms begin proliferating. As they grow, they consume available soil nitrate and give off carbon dioxide. As their population dies off, soil nitrate again becomes available to plants.*

## Animal Manures

Animal manure is often treated as a waste rather than an asset, particularly where the ratio of livestock to cultivated land is too high. As a result, millions of dollars worth of nitrogen, phosphorus, and potassium is discarded annually, and allowed to pollute our rivers, rather than being returned to the soil. Perhaps more important than its nutrient value, manure contributes substantially to soil organic matter and biological activity. It is rich in numbers and diversity of decomposer organisms, thus stimulating the humification process for whatever other organic materials are added to the soil.

**Use of Raw Manure:** Any manure that has not been left to age for at least six months, or has not been put through a composting process, is considered "raw." Raw manure, especially from poultry, is similar to synthetic nitrogen fertilizers in its effect on soil. It may also contain weed seeds, which pass unharmed through animals' digestive systems.

Raw manure should be avoided just before planting any produce or grain crop, and throughout the growing season. This is because it releases more nitrogen than crops need, resulting in poor quality and pest susceptibility. The excess nitrogen also increases weed growth, or is wasted. Raw manure should never be spread on snow or frozen ground: Most nutrients will be lost to the crops, and will contribute to water pollution through runoff instead.

The best time to spread raw manure is when it can be incorporated into the soil without delay. Some time for decomposition should be allowed before planting. As much as one-quarter of the nitrogen content can be lost due to ammonia volatilization in one day, and nearly half after four days.

In climatic zones where animals graze year-round, the use of a chain or pasture harrow is ideal to work the manure evenly into the soil.

**Manure storage:** Methods of storing manure vary in cost and in the effectiveness with which they preserve nutrients. Some are well-known and accepted; others are still experimental. The right one for you depends mainly on available resources and personal inclinations.

In general, solid storage systems (such as sheltered piles) are cheaper and produce less of an unpleasant odor than liquid or slurry systems (such as lagoons). Solid storage systems require more labor to operate, but are cheaper, whereas liquid systems are costly to install but easy to manage. Some farmers have experimented with homemade aerated slurry systems, which offer the ease and convenience of liquids, but reduce odors and stabilize nutrients by aerating the contents through periodic agitation. Commercially available aerated slurry systems are currently very expensive.

Since most of the nutrient loss in manure during storage is caused by leaching, outdoor piles should ideally be covered, even if only with a layer of bedding every few days. Increasing the amount of bedding in the stall will also help retain urine, as well as improve the carbon:nitrogen ratio for more thorough decomposition. If shelter is impossible, water that runs off the pile should be channeled into a holding tank or lagoon where it can be used (after filtering) for irrigation or to dilute foliar nutrient sprays. Many European systems have separate storage facilities for urine and runoff.

Ammonia loss from cow manure can be reduced by adding rock or colloidal phosphate to the gutter at the rate of one to two pounds per cow per day. This is also an excellent way to utilize the otherwise largely unavailable phosphate content from that source. Some farmers prefer to add it when loading the manure spreader, to avoid abrading gutter cleaning equipment.

The same guidelines for spreading raw manure apply to aged manure, especially the need to incorporate it quickly. However, well-aged manure is less likely to upset soil systems with soluble nutrients, and so can be used more safely just before planting.

## Composting

Why compost? Composting is the art and science of mixing various organic materials in a pile, monitoring the resultant biological activity, and controlling conditions so that the original raw substances are transformed into a stable humus. This process of humification is a combination of biochemical degradation and microbial synthesis. Composting is a form of aerobic digestion or controlled fermentation, and differs markedly from anaerobic breakdown or putrefaction. Composting can be done on any scale—garden, farm or industrial.

High quality compost consists primarily of humus and offers the same benefits: it creates and supports the biological processes in the soil. Many people object to the extra time and expense necessary to make compost, asserting that the humification process takes place naturally when organic wastes are incorporated into the soil (referred to as sheet composting). Compost proponents counter that only composting can guarantee humus as the end product, imparting a quality rarely attained in sheet composting.

Moreover, sheet composting will not readily decompose many raw materials.

Very healthy soils are able to buffer the soluble portions of raw manure, break down carbonaceous materials such as sawdust, and digest such wet materials as cannery wastes. However, such soils are the exception. Compost is a microbe-laden substance which inoculates the soil, activating its biological processes. It is a source of organic matter as well as carrying a modest mineral fertilizer value.

In one study that compared compost with stockpiled feedlot manure in the Midwest, researchers found that test plots using compost resulted in yields similar to those fertilized with four times the amount of manure. In addition, measurements of soil quality (pH, organic matter levels, CEC, nitrogen, phosphorus and potassium levels) as well as plant tissue (nitrate, phosphorus and potassium) showed improvement on plots getting compost. These differences cannot be accounted for by the actual nutrient content of the compost, indicating that it probably improves the availability of existing soil nutrients.

Composting does involve extra time and money, but offers the following practical advantages in addition to the benefits described above:

- It stabilizes the volatile nitrogen fraction of manure by fixing it into organic forms (usually microbial bodies).
- It allows use of materials that may be toxic to soil organisms, such as cannery wastes, or that will steal soil nitrogen if applied directly to soil in the raw state, such as sawdust.
- It permits an even distribution of trace mineral soil amendments, avoiding the problem of spot imbalances.
- It eliminates most objectionable odors created by bacterial action on sulfur and nitrogen compounds.
- It reduces the volume of wastes, and therefore the number of trips to the field.
- It eliminates most pathogens and weed seeds by thermal action (when pile temperature reaches 60° C).

- The final product is easy to store and handle, and versatile in its applications.

## On-Farm Waste Recycling

As landfills close and tipping fees increase, composting is becoming a lucrative farm side-line, not so much from selling the finished product as a soil amendment, but as an organic waste recycling enterprise. Many municipalities are happy to deliver yard wastes and leaves directly to local farms free of charge, but frequently industrial processors, supermarkets, and other businesses will pay well for the service of hauling them away. In New Jersey, for example, one farm accepts cranberry wastes from a local plant, charging the company only one-fourth the going cost of landfilling their "waste." This translates into a substantial source of income before realizing anything from sales of the finished compost.

Before deciding to pursue industrial composting on the farm, however, keep a few warnings in mind:

- Some organic wastes can carry toxic contaminants, such as heavy metals, pesticides or industrial by-products such as PCB and dioxin. Some contaminants can be rendered harmless or stabilized by the composting process, but be sure to demand a laboratory analysis of any raw wastes you consider accepting. If the farm is to be certified, consult your certification agency regarding the acceptability of any off-farm wastes you want to use.
- You will probably need a permit to operate as a waste disposal site, which often presents overwhelming barriers when neighbors conjure up images of community health hazards and unpleasant smells.
- Depending on the scale, some capital investment in composting equipment, shredder, screener, and perhaps bagger will be necessary.
- If you intend to market the finished compost, you have to be prepared to give as much attention to marketing and quality control as you would any other farm enterprise.

- Most industrial organic wastes must be mixed with some other ingredient to achieve the proper C:N ratios needed for successful composting. Often this will require locating and purchasing large quantities of carbonaceous materials, since the most problematic wastes are usually high nitrogen, wet, sloppy, and smelly.

## Composting Equipment & Methods

Compost making on a garden scale has been ably described elsewhere. Optimum pile size, or critical mass, is about five feet square and four feet high, or close to two tons. The only equipment needed are manure forks and other loading tools. Many garden catalogs offer specialized compost bins, tumblers, and aerating tools in addition to shredding machines. Shredders do reduce the size of raw materials and produce more homogeneous piles, but are expensive. Rear-end rototillers can be used to turn large garden compost windrows.

Farm-scale composting requires a small tractor with a power take-off (PTO), a PTO manure spreader, and a second tractor with a front end loader. Most manure spreaders make good compost machines, although side-delivery types and those with unconventional beaters are not appropriate. In regions where manure spreaders are not available, front-end loaders, preferably with buckets that have long teeth, will have to be used.

The procedure is straightforward, but takes practice to master. Piles of material are arranged parallel to the projected length of the windrow. The full manure spreader begins unloading, while the front end loader alternates between the raw material piles, dumping the proper proportion of each into the spreader. Once the pile is about four feet high and six to eight feet wide (with some hand shaping in the early stages), the spreader is inched forward to let the raw materials hit against the new pile and fall into place. As the manure spreader slowly pulls forward, the pile is formed (see Figure 17).

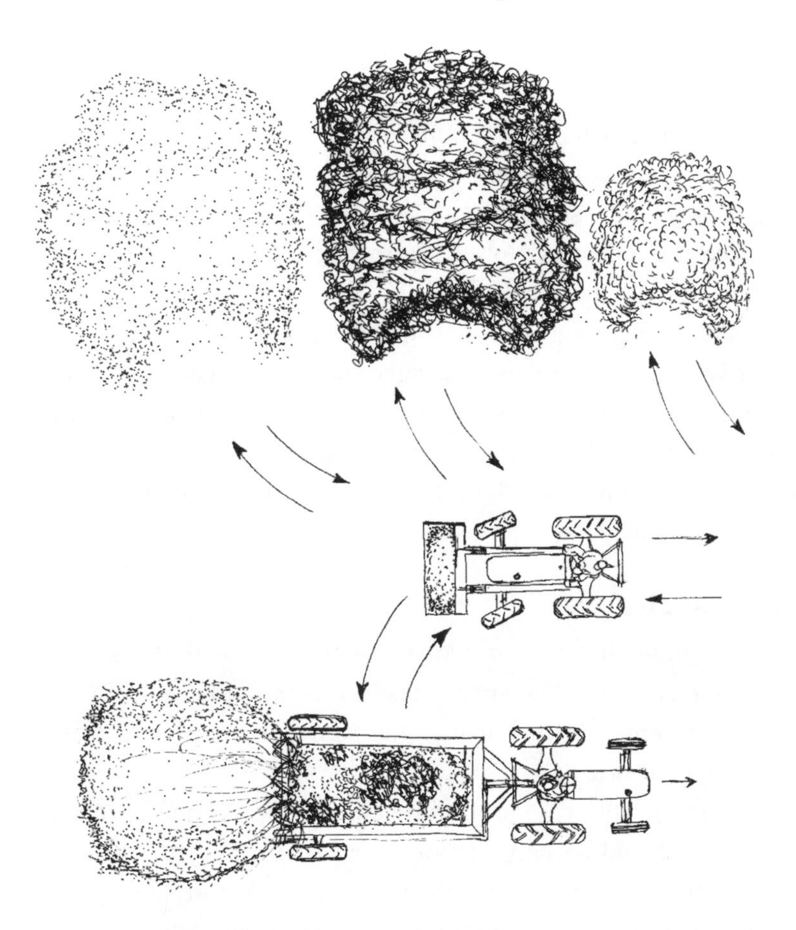

*Figure 17. Farm-scale composting technique. Layout of piles of raw materials relative to windrow placement. Front-end loader alternately dumps materials into continuously operating manure spreader. (Drawing by Stewart Hoyt.)*

The manure spreader not only mixes the materials, but aerates them as it chops and throws them. This is quite different from using a front end loader to dump (compact) materials into a windrow shape. While this is not an unnatural job for the manure spreader, it is intensive, so lots of grease and oil should be used on the chains and gears. Depending on condition of equipment and personal comfort, you can compost at any time of year.

Climatic conditions will also influence the optimum pile size and ratio of materials.

Compost manufacturers, mushroom growers, research institutions, municipal sludge plants, and some farmers have developed machines specifically for compost production. These machines are designed primarily to turn piles of raw materials. Some straddle the windrow, while others turn it from the side. Turning, which controls air and moisture levels, increases the speed of composting. These machines may be equipped with tanks to spray water, which often contains bacterial inoculants. Some intermediate scale machines, such as the "Easy Over" compost turner, can be powered by a farm tractor. Such machines could make custom composting a viable option in the near future.

## Raw Materials

Anything that was once alive can be composted; it is simply a matter of creating an appropriate initial mix. Materials can be judged according to their carbon:nitrogen ratio, moisture content, and physical structure.

A reasonable carbon to nitrogen ratio for the initial mix is between 25 and 35 parts carbon to one part nitrogen. This ratio allows the nitrogen to "fire" the mix and decompose the carbonaceous material. A higher ratio (more carbon) will take longer to compost, slowly exhaling carbon dioxide until the proper ratio is achieved. A lower ratio results in higher nitrogen loss through volatilization or leaching.

Manure is ideal material for composting. If it must be stored for any length of time before composting, it should be mixed with bedding or litter, creating a superb base material. For example, when mixed with bedding, dairy manure has an average carbon:nitrogen ratio of 25, although usually it is too wet. Organic materials rich in nitrogen which may be available in different areas include slaughterhouse by-products, fish wastes, leather tankage, and peanut, sugar cane and cocoa wastes. They can be mixed with carbonaceous materials such as hay, straw, sawmill waste, vegetable wastes, bagasse, dry leaves, clippings, and rice hulls. In many tropical zones, where compost is especially valuable, many raw materials such as saw chips, sugarcane bagasse and filter-mud, rice hulls, and other wastes from large-scale export crops are readily available (see Table 12).

Nitrogenous materials are often wet, and can be balanced by drier carbonaceous wastes. If the mixture is too wet, air is excluded; too little water will not support microbial needs. The optimum initial moisture level is around 60%. The mix can be tested by squeezing a handful until the particles stick together. If it drips, it is too wet. If it falls apart when released, it is too dry.

Particle size and shape are also important for composting. For example, hay and straw have a lower carbon:nitrogen ratio than sawdust, but may be less useful for composting because they absorb water only at their ends, and tend to mat, excluding air, even when put through beaters. Sawdust (actually the small chunks from a circular saw blade, not dust from a sander or shavings from a planer), because of its granular texture, helps form a crumb structure in a windrow when mixed with manure. This creates good aeration and an end product that is crumbly and easy to spread.

Rock powders such as rock or colloidal phosphate, granite dust, and basalt meal can be added to the mix. They aid microbial growth in the pile, and the composting process makes them more available to plants. Lime should not be mixed with raw manure, as it liberates ammonia. Conventional soluble fertilizers tend to inhibit microbial processes, and should be avoided. Do not use sewage sludge and night soil in composts intended for use on human food-producing land, because of possible pathogen and heavy metal contamination. They can be used in composts for shelter belts and other non-food crops, and possibly green manure and forage crops. Soil is a valuable addition to initial mixes, often contributing around 2–5% of the content when scooped up

*Table 12. Listed below are approximate nutrient compositions of materials that are commonly available in the Northeast. Except for manures, information is provided only for nitrogen (N), Phosphate ($P_2O_5$) and potash ($K_2O$), though all of these materials contain other essential macro- and micronutrients. Carbon:nitrogen ratios are indicated for bulky materials that may be used for compost.* ***(A)*** *High analysis organic wastes: Carbon:nitrogen ratio generally 24:1 or less (Often commercially available.)* ***(B)*** *N-P-K, organic matter, and moisture content of various manures (%).* ***(C)*** *Carbon:Nitrogen ratios of bulky organic materials.*

**(A)**

| Material: | Percent composition N | $P_2O_5$ | $K_2O$ |
|---|---|---|---|
| Alfalfa leaf meal | 3.0 | 0.3 | 0.0 |
| Apple pomace | .20 | .02 | .15 |
| Blood meal | 8.0-13.0 | 1.5-2.0 | .6-1.0 |
| Bone meal, steamed | .70-4.0 | 18.0-34.0 | —— |
| Cottonseed meal | 6.0 | 2.5-3.0 | 1.0-1.7 |
| Feather meal | 12.0 | —— | —— |
| Fish meal | 10.0 | 4.0-6.0 | —— |
| Hoof & horn meal | 12.5 | 1.75 | —— |
| Leather meal | 5.50-12.0 | —— | —— |
| Seaweed (kelp) | 0.6-0.9 | 0-0.5 | 4-13 |
| Sewage sludge | 1.5-2.0 | 1.4 | .4-18 |
| Sewage sludge, activated dry | 2.0-6.0 | 3.0-7.0 | 0-1.0 |
| Soybean meal | 6.9-7.0 | 1.2-1.6 | 1.5-2.3 |
| Tankage, animal | 6.0-9.0 | 6.0-10.0 | 0-1.5 |

**(B)**

| Kind of Manure | Nitrogen | Phosphate | Potash | Organic Matter | Moisture |
|---|---|---|---|---|---|
| Rabbit | 2.4 | 1.4 | 0.6 | 33 | 43 |
| Chicken | 1.1 | 0.8 | 0.5 | 25-45 | 55-75 |
| Sheep | 0.7 | 0.3 | 0.9 | 32-34 | 66-68 |
| Horse | 0.7 | 0.3 | 0.6 | 22-26 | 74-78 |
| Steer | 0.7 | 0.3 | 0.4 | 17 | 83 |
| Cattle | 0.6 | .02 | 0.5 | 17 | 83 |
| Duck | 0.6 | 1.4 | 0.5 | 25-45 | 55-75 |
| Pig | 0.5 | 0.3 | 0.5 | 14 | 86 |

(C)

| Material | Ratio |
|---|---|
| Vegetable wastes | 12:1 |
| Alfalfa hay | 13:1 |
| Seaweed | 19.1 |
| Rotted manure | 20:1 |
| Apple pomace | 21:1 |
| Legume shells (peas, soybeans, etc.) | 30:1 |
| Leaves | 40-80:1 |
| Sugarcane trash | 50:1 |
| Cornstalks | 60:1 |
| Oat straw | 74:1 |
| Chaff & hulls (various grains) | 80:1 |
| Straw | 80:1 |
| Timothy hay | 80:1 |
| Paper | 170:1 |
| Sugarcane fiber (bagasse) | 200:1 |
| Sawdust | 400:1 |

with the raw materials. It also helps inoculate the pile with "native" microorganisms. For best results, use a wide variety of materials, including animal manures.

There are many compost activators on the market. Those with dormant bacterial spores are useful for composts made without cow manure or other materials rich in microbial life. Some so-called compost activators consist mainly of nitrogen fertilizers. Bio-dynamic preparations are not activators, per se, but specific aids and regulators of the humus-plant-cosmos energy flow.

## Monitoring & Analysis

The activity of a compost pile can be divided into a break-down and buildup phase. The microbial activity begins in a new pile almost immediately. As the microbe population explodes, it generates heat. Within three to five days the pile will reach a temperature peak of 45° to 55° C. The pile should not be allowed to exceed 60° C, as mineralization will occur. As it slowly cools, the organic matter broken down by one group of organisms is

*Table 13. Trouble-shooting guide to composting.*

| Observation | Possible Problem | Corrective Action | |
|---|---|---|---|
| | | **Short-term** | **Long-term** |
| Pile does not heat up. | Not enough nitrogen. | Add nitrogen by mixing in fish, blood meal or manure. | Lab test raw materials for C:N ratio. |
| | Too wet for aerobic bacteria. | Turn pile and mix in dry carbonaceous material. | Balance pile moisture at 60% average. |
| | No air penetration. | Turn pile to aerate. | Shred material. Add a granular material. Upgrade turning equipment. |
| Pile heats slowly. | Low on nitrogen. | Add nitrogen. | |
| | Not enough air. | Turn pile. | |
| | Lack proper bacteria. | Add BioActivator. | |
| Pile overheats and dries out. | Too much aeration. | Pack pile. Add water. | Increase pile density or irrigate pile. Pre-digest manures in low windrows. |
| Pile smells of ammonia. | Too much nitrogen. | Add carbonaceous material. Mix pile more thoroughly. | Change recipe or turning procedure. |
| | Too high pH. | Add sulfur. | Use low pH materials. |
| Pile leaches after rain. | Too moist to absorb additional water. | Turn pile. Use drier materials. Cover pile. | |
| Anaerobic core in pile. | Uneven pile mix. | Turn pile. | Upgrade turning equipment or procedure. |
| | Materials that mat. | Turn pile. | Shred materials or switch to granular materials. |
| Weeds on pile. | Infrequent turning. | Turn pile. | Increase turning frequency. Cover with new material. |

transformed into humus by another group. When composting is complete, the original mass of raw materials will be reduced by about half, the rest having been transformed into carbon dioxide by microbes.

If the carbon:nitrogen ratio, moisture content, particle size and contours of the pile are ideal, there is no further need to manipulate the pile. However if the optimal conditions are not present, then some action may be necessary. If the pile is too wet, an anaerobic core will form, with little temperature increase. The pile must then be turned. If there is not enough moisture, or if microbial heat has driven off the moisture, water must be pumped in or the pile must be packed or spread out.

On a farm scale, the process of reloading the manure spreader and throwing a new parallel pile is much quicker than the first mixing of raw materials. Industrial compost turners mix their piles every few days, as their purpose is to generate a homogenized compost in as little time as possible.

Decisions about whether and when to turn compost can be based on temperature, monitored by long (two or three feet) thermometers, or by feel. A composter also develops a fine sense of smell—finished compost has an easily recognized earthy fragrance. Anaerobic cores or the grey mold from overheating can be seen when a pile is opened with a spade. Lab analysis of the raw material mix and at various stages of the process may be helpful, but often the senses of the composter can best direct the composting process.

## Compost Usage

High quality finished compost can be spread at any time in any amount, limited mainly by equipment capabilities. The greatest benefit from its inoculating effect is gained by spreading available compost thinly over the total acreage, rather than concentrating on a small area. Differences in raw materials and extent of processing often determine differing uses. Semi-finished compost contains more soluble nutrients, especially nitrogen, and is often used for heavy feeders such as corn. However, the intermediate breakdown products in partially finished compost can be detrimental to germination of some crops. If minerals are needed, they can be blended into the compost at the last turning, or used to create "boosted blends" for special needs like greenhouse media.

Compost fosters the biological processes in the soil. Its use is a major tool in the creation and preservation of soil fertility.

## Green Manures & Other Soil-Improving Crops

Among the most important ways of building and maintaining soil organic matter is to grow it in place. The obvious benefit of this method is that there is no need to load, haul, and spread large quantities of materials. All you have to do is spread some seed and perhaps add some supplemental nutrients, and the biomass will accumulate as the plants grow, right where it's needed. There are drawbacks, including having to take land out of cash crop production for a period of time and concerns about competition for moisture, but the effort invested in crop planning and perhaps some on-farm trials of different green manures, will be amply rewarded in improved tilth and humus content. Recent research indicates that, in areas of low rainfall, subsoil moisture removed by green manure crops is more than offset by improvements gained in the soil's moisture-retention qualities and overall tilth.

Green manures can also be important sources of biological diversity, crucial for ecological pest control systems. Beyond serving as smother crops, they interrupt pest and disease cycles, and provide a habitat and food source for beneficial insects such as predatory mites, wasps and spiders. The down side is that crop pests such as tarnished plant bugs may also be harbored in adjacent green manure plantings.

A green manure can be defined as any crop that is tilled in before maturity to improve the soil. It serves mainly as a source of nitrogen (especially legumes) and possibly some organic matter, but may also contribute significant amounts of other essential nutrients. When allowed to reach maturity before incorporation, soil improving crops may also contribute significantly to soil organic matter. Green manure root systems bring otherwise unavailable nutrients from the subsoil to the surface, and prevent erosion on cultivated land between cropping seasons. The greatest contribution to soil organic matter will be made by those crops which produce the most biomass—usually perennial grass or legume sod crops, which have massive root systems.

Other terms may also be applied to crops grown primarily to benefit the soil, depending on their intended purpose. Terms such as cover crop, smother crop, catch crop, intercrop, and living mulch reflect different ways to use soil-improving crops in farm planning.

Cover crops are used to protect the soil from erosion during the dormant season, and may also help smother weeds, in which case they may be called smother crops. They may be undersown or planted after a main crop is harvested, then tilled under at the beginning of the following growing season. Although they usually provide a flush of nutrients as they decompose, cover crops rarely contribute to stable humus formation. Winter rye and oil radish are commonly used fall cover crops in the North, but some farmers prefer crops such as oats or millet, which die over winter and are easy to incorporate in the spring. Southern farmers use drought-tolerant peas and other annual legumes as summer cover crops.

In wind-swept prairie regions, crops planted to trap snow and improve soil moisture between main crops are called catch crops. Crop residues or "trash" left on the surface over winter can serve the same end. In humid regions, a catch crop may be planted to prevent nutrients from leaching away.

Rotation plans often include perennial sod crops, grown for two years or more as forage, which become green manure when plowed under in preparation for the next crop. For greater soil benefit, the final cutting of hay may be left in the field. Annuals such as buckwheat and millet are also commonly used as a main season green manure or smother crop in rotation plans. These and other fast-growing species may also serve to protect the soil in preparation for transplants or late season crops. Annual weeds such as lambsquarters and mustard can be used similarly, as long as they are not allowed to form seed.

A green manure planted at the same time as the main crop, between the rows or in alternating strips, is called an intercrop or living mulch. In temperate regions a low growing perennial, often clover, can be sown between rows of annual crops. Besides protecting soil that would otherwise be exposed, it adds nitrogen as it grows. Grasses such as annual rye may be mown, and the clippings provide a free mulch to the main crop. Strip cropping of sloping lands, with alternating sod and cultivated crops, is another form of intercropping.

Mulches—living or dead—serve a critical function in tropical soils, reducing soil temperature as well as contributing fresh organic matter. Alley cropping or agroforestry systems may use perennial legumes, including shrubs and trees, between rows of annual or perennial cash crops. Besides the nitrogen contributed by the legume root systems, the tops may be cut for fodder or as mulch for the main crop, or used to shade crops such as coffee and cocoa. Other types of intercropped plants, such as sunhemp *(Crotalaria ochroleuca),* can serve as trap crops to attract pest insects away from the main crop.

## Selection, Timing & Management of Green Manures

Just about any plant—including weeds—can be a green manure. The most important considerations are selection of the most appropriate variety for your needs, and timing.

Table 14 lists characteristics of the better known soil-improving crops. You may be able to get more information about locally adapted species and varieties from Extension or other agricultural agencies. Factors to consider in selecting a green manure include its primary purpose, its adaptability to your soil type and microclimate, its tolerance for adverse conditions, the characteristics of your main crop, and of course, the availability and cost of seed. The appendices list resources for evaluating and locating sources of green manure varieties.

Synergistic combinations of different green manures are also possible—a grass and legume are commonly sown together, enhancing the growth of both. Mixtures that include drought-tolerant species with more demanding ones help spread the risk of poor germination in a dry year. Evaluation of green manures is one area in which on-farm experimentation has become the cutting edge of practical new information on fertility improvement.

Timing is critical, both for sowing and incorporating green manures. Growth cycles must be timed to complement, not compete with, cash crops. Some will grow happily in cold soils, and others can tolerate midsummer conditions. Biennials and perennials produce the most biomass in their second or third year of growth. As the green manure crop grows, its nutrient composition and carbon:nitrogen ratio changes. The trick is to pick the right time to incorporate it, and thereby provide the most benefit to the soil for your effort.

Some characteristics of green manure crops at different stages of growth include:

**Early:** Young, succulent growth decomposes quickly (with good moisture and aeration), and releases mineralized nutrients for the crops that follow. Carbon:nitrogen ratios are low, so microbes lock up little of the nitrogen, leaving it immediately available to crops. Decomposition is fairly thorough, with little contribution to stable humus.

*Table 14. Green manures.*

| Type | When to sow | Seeding rate (lbs per 1,000 ft$^2$) | Where adapted in US | Comments |
|---|---|---|---|---|
| LEGUMES | | | | |
| Alfalfa | Spring | 1 | all | Deep-rooted, needs good drainage, pH. |
| Beans, soy | Spring | 5 | all | Annual. |
| Beans, fava | Spring or Fall | 6 | all | Annual, edible bean, cold-tolerant. |
| Clover, alsike | Spring-late summer | 1/2 | North | Tolerates wet, acid soil. |
| Clover, crimson | Fall | 1 | South-Central | Winter annual. |
| Clover, ladino | Spring-late summer | 1/2 | all | Tolerates traffic, wet & droughty soil. |
| Clover, red | Spring-late summer | 1/2 | North-Central | Perennial, good P-accumulator. |
| Clover, sweet (yellow or white) | Spring-summer | 1/2 | all | Needs good drainage & neutral pH. Yellow tolerates dry conditions. |
| Clover, white dutch | Spring-summer | 1/2 | all | Perennial, tolerates traffic. |
| Cowpeas | Spring | 5 | South-Central | Drought-resistant annual. |
| Hairy Indigo | Spring | 1/2 | Deep South | Needs warm, well-drained soil. Resists root-knot nematode. |
| Lespedeza | Spring | 1 | South | Good for restoring eroded, acid soil. |
| Lupine, white/blue | Spring | 1 | all | Tender annual, good OM producer. |
| Pea, field or Austrian | Spring or fall | 5 | all | Annual, best combined with grain crop. |
| Trefoil, birdsfoot | Spring | 1/2 | all | Comparable to alfalfa but tolerates poor soil. |
| Vetch, hairy | Spring-fall | 1.5 | all | Best combined with rye; good biomass producer; winter hardy. |

| Type | When to sow | Seeding rate (lbs per 1,000 ft$^2$) | Where adapted in US | Comments |
|---|---|---|---|---|
| NONLEGUMES | | | | |
| Barley | Spring or fall | 2.5 | all | Needs pH 7-8; use spring varieties in North. |
| Brassicas (kale, radish, etc) | Spring-fall | 1/2 | all | Fast-growing, cool season. Do not allow to set seed. |
| Bromegrass, smooth | Fall | 1 | North | Cold-hardy winter cover crop. |
| Buckwheat | Spring-summer | 2-3 | all | Tender, good smother crop; P-accumulator. |
| Millet, pearl | Spring-summer | 1 | all | Fast-growing warm season smother crop. Tolerates low pH. |
| Oats | Spring or fall | 2.5 | all | Tolerates wide pH range, avoid heavy clay. Good "nurse crop" for legumes. |
| Rye, winter | Fall | 2.5 | all | Allelopathic (suppresses weed growth when turned under). |
| Ryegrass, annual | Spring or fall | 1-2 | all | Widely adapted, rapid grower. |
| Sudangrass (and sorghum) | Spring-summer | 1 | all | Tolerates poor drainage; rapid biomass producer in hot weather. |
| Wheat, winter | Late summer | 2-3 | all | Needs pH 7-8 & good fertility. |

TROPICAL GREEN MANURE CROPS, grown under widely varying soil, weather and seasonal conditions, can include the following: *Aeschynomene, Acacia saman, Azolla, Canavelia, Chromalaena odoratum, Crotalaria, Indigofera, Gliricidia sepium,* grain legumes such as mung beans, *Mucuna, Pithecellobium dulce,* and *Sesbania rostrata.*

*Table 15. Nitrogen use.*

*(A) General crop nitrogen requirements.*

| Crop | Pounds of nitrogen per acre per year |
|---|---|
| Grass (2-3 times as a top dressing) | 100–150 (maximum) |
| Small grains | 20–40 |
| Potatoes | 120–160 |
| Leafy vegetables | 120 |
| Root crops | 80 |
| General home vegetables | 100 |

*(B) Nitrogen contributions of green manures & sod in rotation.*

**Highest: (over 100 pounds per acre)**

LEGUMES

- Perennial & biennial (alfalfa, trefoil, red & sweet clover, vetch)
- Turned under as green manure
- Sod after hay is harvested*
- Annual (soybeans, field peas) turned under as green manure

**Moderate: (50–100 pounds per acre)**

- Sod of mixed grass and legume in rotation*
- Non-legume annuals turned under as green manure

**Low: (under 50 pounds per acre)**

- Sod of perennial grass only
- Residue after harvest of annual legume crops (soybeans, dry beans, field peas)

*Weight of roots relative to tops increases with age of perennial stand and is highest in early summer of second year for biennials.

**Midseason:** Growth is coarser and may require more powerful cultivating equipment to incorporate. Decomposition is slower, and may not provide surplus nutrients to the crops that follow. Carbon:nitrogen ratios favor maximum microbial growth, and the nutrients they use will become available later in the season. Larger amounts of organic matter are added to the soil, improving structure and tilth.

**Late:** Beekeepers may wish to use flowering green manures for bee forage. Otherwise, green manures should not be allowed to form seed; if they do, they are no longer "green" manures. On poorly drained soils, waiting as long as possible will improve aeration the most; this may be the best time to work the soil. High carbon:nitrogen ratios result in very slow decomposition, with nutrients well protected against leaching.

Whatever its stage of maturity, a green manure is most effective when mixed into the top few inches of soil, rather than being plowed under where air supply for microbes is limited. It is best to incorporate green manure in two passes, allowing it to partially dry before final incorporation. This prevents putrefaction caused by excess moisture. Mineral fertilizers and rock powders are often most effective when incorporated with green manures. If spread prior to seeding they will enhance the growth of the green manure, increasing the yield of organic matter. Raw manure can be incorporated with overly mature, carbonaceous green manures, speeding decomposition and humus formation.

## Rotations

The only circumstance in which rotations are not necessary in an ecological soil management program is in the case of perennial plantings. Even then, a polycultural system using rotated annual intercrops is often valuable. The rotation plan is a key element of your overall farm planning system, which is central to compliance with organic certification requirements.

Whenever cultivated annual crops of any kind are grown, rotation, succession or intercropping with sod or other perennial crops is necessary, unless large amounts of organic matter are brought in. Rotation of plant types enhances soil fertility by mixing nitrogen-fixing legumes, mineral accumulating or scavenging plants, and deep-rooting species. Rotations regenerate organic matter exhausted by previous crops by adding crops that maximize biomass or provide sod cover. Rotations also perform the indispensable function of controlling weeds by smothering or allelopathic effects. They also contribute to the control of crop pests without the use of poisons by providing the ecological diversity required by predators and parasites. This biological diversity is matched by a diversity of labor requirements and farm income sources.

## Rotation Planning Considerations

**Length of rotation:** Analyze conveniently divided portions of fields. Each area suited to the same general crops can be put into the same rotation pattern. If you have as many portions as there are years in your rotation, a good balance of high and low income crops can be maintained. In general, longer rotations promote greater ecological complexity, and therefore greater stability. If possible, divide large fields into smaller rotation units.

**Soil characteristics:** "Native" organic matter content and fertility, as well as texture, drainage and slope, will limit what crops are suited to your land. Your options will increase as your ecological management program progresses. Matching crops to soils is an art developed through practice. This is where the experience of neighbors and local agricultural advisors can be most useful. In general, highly fertile, flat land can be cropped most intensively; that is, less of the rotation can be devoted to sod crops and more to crops with high yields and returns. Poor soils may require longer periods in soil-improving crops, until fertility levels can support more intensive production, or may never be suitable for cultivated crops.

**Crop characteristics:** The more demanding a crop, the less frequently it should be grown in the same place. Crops which are susceptible to the same soil-borne insects and diseases must be moved around to prevent pest buildups. Forage crops help clean a field of annual weeds, but fine-seeded, slow germinating crops like carrots have trouble following freshly tilled sod. Alternating deep-rooted and shallow-rooted crops makes subsoil nutrients more available and improves soil structure. Where rainfall is limited, water requirements of succeeding crops must be factored in. Some rotation schemes call for alternating root crops, fruit crops, and soil-improving crops.

Crops that have particularly heavy demands for one nutrient should be followed by those that need comparatively little of that same nutrient, while soil-improving crops should be timed to precede heavy feeders. The common practice of following alfalfa with corn in a rotation can usually provide all the corn's high nitrogen needs. In an ideal situation, cultivated row crops should not be grown more than two years in succession and should include a green manure cover between cropping seasons. (Table 16 lists nutrient demands of various crops.)

**Rotation effect:** Some evidence indicates that certain crops have a beneficial effect on succeeding ones, while other combinations may be detrimental. Brassicae seem to do better following onions, and potatoes following corn, while potatoes seem to be more prone to scab when they follow peas or oats. Cotton following corn has been shown to produce higher yields than either crop alone.

**Other sources of fertility:** As long as any crop or livestock product is sold off the farm, there is little chance of meeting all soil needs through rotations, even when everything else is recycled conscientiously. If your proportion of livestock to cultivated land is high, organic matter levels can be maintained primarily through manure. If you purchase livestock feed, you are still buying fertility from off-farm. Other sources of organic matter, such as mulches, may also reduce the need for sod in rotation. Timing of manure, compost and mineral fertilizer applications should be built into rotation plans, to coincide with crop requirements and optimization of green manures.

**Seasonal labor needs & market availability:** Rotations allow you to spread work loads more evenly over the growing season,

*Table 16. Nutrient demands of various crops.*

| Crop | Nitrogen demands | Phosphorus demands | Potassium demands | Other critical nutrients |
|---|---|---|---|---|
| FORAGES | | | | |
| Grass | Mod-High | Moderate | Mod-High | magnesium, iron |
| Legume | Low | Low-Moderate | High | calcium, molybdenum, boron |
| Small Grains | Low-Moderate | Low–Moderate | Low-Moderate | magnesium, iron |
| Corn | High | Low-Moderate | High | zinc, magnesium |
| Potatoes | High | High | High | manganese |
| Root Crops | Low | Low-Moderate | Low-Moderate | boron |
| GARDEN VEGETABLES | | | | |
| Beans & peas | Low | Moderate | Low-Moderate | calcium, molybdenum, sulfur, zinc |
| Brassicae (cabbage family) | High | Mod-High | Mod-High | iron, boron, calcium |
| Cucurbitae (squash, cukes, etc.) | Mod-High | Moderate | Moderate | magnesium |
| Leaf crops | High | Moderate | Moderate | iron, copper |
| Onions | Mod-High | Moderate | Low-Moderate | copper, sulfur |
| Tomatoes (peppers, eggplant) | Moderate | Low-Moderate | Mod-High | iron, magnesium |

as long as you plan properly to avoid several crops whose peak labor demands coincide. Specialized equipment may also be needed to plant or harvest certain crops. Crops which work well in a rotation may also be difficult to market, requiring you to factor in a low return per acre or to invest effort in developing new markets. However, the financial risks of crop failure or market fluctuations which may result from reliance on a single crop are also diminished.

## Sample Rotation Plans

**Temperate Zone Vegetables:**

Mechanized, moderate-scale (Lampkin): Onions-Potatoes-Carrots-Brassicae-Legume forage. This presumes use of manure or compost, and cover crops between seasons (e.g., winter rye after potatoes, and/or buckwheat before carrots).

Small-scale, intensive (Coleman): Legume/green manure-Strawberries (2 years)-Carrots-Peas-Brassicae. Also presumes use of manure or compost, and cover crops between carrots-peas and peas-brassicae.

**Cash grain & livestock:**

Temperate, moderate moisture: Alfalfa or mixed legume sod (2 years)-Corn-Soybeans-Oats or other small grain (serves as nurse crop for forage seeding). An additional year of corn can be included if manure or compost is used. Corn is assumed to be harvested for grain, with stubble left in the field.

**Arid Regions:** The overriding concern of arid region farmers when planning a rotation is moisture. They fully understand that the best way to improve the moisture retention of the soil is to add organic matter but they are reluctant to put a "brown manure" crop in the rotation because of the water it will use in growing. They know from experience that the next cash grain crop will be light unless they receive more than normal rainfall. The tendency then is to summer fallow a field after a grain crop. Summer fallowing is simply cultivating a field to keep it bare of moisture-consuming weeds. This practice also lowers soil organic matter and hence the health and moisture-retention abilities of the soil. Farmers can only pull out of this vicious downward spiral by developing a rotation that increases soil organic matter content. The real issue is not agronomics but economics, that is, long term sustainability versus short term financial survival.

It would be close to impossible to accomplish this change all at once but it can be done field by field over a period of years. One of the agronomic keys to this transition is cash crop diversification where some wheat or other grain is replaced by oilseeds such as canola or safflower and dry fallow periods are used to grow pulse crops that don't use much water. The return of animal stock to prairie farms is another important facet in establishing sound rotations. Legumes such as sweet clover (depending on water availability) and black medic can also find a place in the rotation, either seeded under grain or sunflower crops, or as strips between grain (left standing they can catch snow in the winter in northern areas). There are numerous moisture retention tricks developed by organic dryland farmers such as the Australian Keyline system, the development of chickpea and red lentil crops in Turkey or pea varieties in Saskatchewan, alternate low/high strip cutting of grain, the planting of windbreaks to reduce evapotranspiration, and stubble mulch tillage practices.

Each farm will have its own optimum rotation but the basic principles for ecological soil management in dryland areas are:

- Minimize use of summer fallow.
- Minimize tillage.
- Maximize bio-mass incorporation.
- Add manure and carbon sources when possible.
- Put as much area under sod as possible.

The economic key to this conversion is a shift in government policy from current crop base commodity programs to increased support for crop diversification incentives and conservation reserve programs.

**Tropical crop rotations:** Tropical crop rotations are determined according to the rainy and dry seasons, which vary widely in intensity, duration, timing, and reliability. The goal is to keep the soil covered continuously, timing plantings to take advantage of moisture for germination and to avoid erosion of exposed soil. Legumes are a desirable component of the rotation scheme due to their soil-building qualities, although many farmers feel they cannot afford to return them to the soil. In some situations two to four crops, including corn, peanuts, yams, millet, beans or other legumes, and a cash crop, are grown together in the same field. This, along with techniques like alley cropping, mulching, and livestock grazing between cropping periods, lessens the need for the classical type of crop rotation.

**Permaculture:** While rotations may be appropriate for many annual crops in some climatic regions, a permacultural system may be a better ecological design for perennial crops and other climatic regions.

Permaculture is based on the design of cultivated ecosystems modeled on the diversity, stability, and resilience of natural ecosystems. Permaculture utilizes the principles of agroforestry, alley-cropping, intercropping, and other polycultural systems. This approach often incorporates a multi-story design where tall trees are spaced between short trees and shrubs mixed with annual plants over a ground cover. As with other ecological soil management, permaculture is a synthesis of traditional knowledge and scientific technologies. This information is used in the design process, where plants, animals and minerals are selected based on site specific considerations and available managerial and financial resources. The resulting agricultural configuration has built-in feedback loops that can monitored, evaluated, and modified.

In temperate zone orchards many permaculture design techniques can be used to increase the ecological diversity, thereby enlivening the soil and creating internal, biological pest controls. There is also a set of permacultural practices for arid areas that can slowly increase the moisture retention of the soil/plant system.

Although a permacultural system can be designed for any soil and climatic system, the best examples are currently in the tropics. Organic matter in the soil is continually dissipating in the high temperature tropics. Thus tropical soils cleared of the natural forest or savanna can quickly lose their productivity. Commercial tropical agriculture has often been implemented with large-scale temperate climate zone methods, specifically monoculture row-cropping with intensive cultivation by high-powered tractors. This style of agriculture is not sustainable in either economic or ecological terms, not to mention its negative social impact. In contrast, many profitable tropical organic farms have developed systems that supply food for local consumption, export crops for cash, and improve the soil.

In a Mexican community on the Pacific Coast, sesame is grown as an alley crop between rows of leguminous tamarind trees. Nearby, papaya plants are intercropped in a young mango orchard. Plots of bananas are nestled under coconut trees, while cows and pigs are rotated through various fields. The increasing use of leguminous shrubs, herbs, fruits and other plants, both traditional and exotic, is taking place on an experimental basis in various hedgerows and shelterbelts. This integrated system protects soil from erosion, increases organic matter content, supplies nitrogen, and has a greater marketable yield at lower cost than row-crop monoculture.

On the thin soils of the mountainous region of Darjeeling, in India, the traditional tea plantations are faced with massive erosion problems. The causes of this are typical of many regions: trees are being cut for firewood and even the prunings from the tea plants are burned. There is no return of organic matter to the soil or protection for the forest. Landslides are common and the valleys now flood with every monsoon. Contrast this with the organic tea estate of Makaibari, where two generations of management have created a permaculture model. Tea is the cash crop and loving care is taken of the plants from the nursery to the field, and through the hand-picked harvest to the diligent fermentation and firing at the estate processing center. There are five villages on the farm and it is among the people there that the system achieves ecological and social legitimacy. Workers on Darjeeling tea estates have firm legal rights to employment and housing. All management programs must receive the tacit endorsement of the workers through their union. The estate workers even have a right to the tea prunings for fuel.

In order to implement a permacultural system, management had to make it work for everyone, which is after all the real measuring stick of an organic or sustainable system. The transition started with the installation of pilot bio-gas digestors which provided methane cooking gas from manure. Cash was provided for the purchase of animals to create the manure supply. Repayment started with the sale of dairy, egg, and meat products. Interest was the leftover sludge from methane production mixed with waste materials in village compost sites.

Fodder for the animals was provided on estate land, thus getting the support (or buy-in) for green-manure planting schemes. This planting included leguminous trees, shrubs, and clovers as well as quick growth plants like guatemala grass which was also used to provide mulch (termed "thatch" in tea regions) for the tea plants. While the guatemala grass was grown in special areas, the legumes were integrated into the tea plantings. Until recently, tea production has had a tradition of using leguminous trees for shade. They also break the evapotranspiration loss, and provide erosion control, nitrogen from fixation, and organic matter from leaf fall. These trees form the top story of a four story system.

The third story consists of leguminous shrub species such as *Glyricidia,* indigo, *Crotalaria,* and *Medolia.* These shrubs can be severely pruned at the beginning of the rainy season to allow air flow and then allowed to releaf in the dry season to provide shade. The prunings are allowed to decompose on the soil surface or provide animal fodder. Neem trees also occupy this story. Their fruit and leaves are harvested to produce an extract sprayed on tea plants to ward off the tea mosquito bug and mites. The extract is also used by everyone as a germicidal wash, toothpaste and general tonic. Other herbal preparations are used as plant sprays and village medicines in the Indian Ayurvedic tradition.

The tea plants receive a compost mulch at the beginning of growing season after pruning. At Makaibari the tea prunings are left to decay under the plant in a carpet of low-growing clovers. On the hills and in the valleys the forests are protected from cutting and allowed to grow with only the occasional harvest of leaf litter to provide mulch for nearby new tea plantings. The forest is also home to an incredible variety of insects and animals which prevent any pest epidemics.

It's a pleasant place for tea workers to live, with the family income supplemented by egg and milk sales to local merchants, a convenient gas cooking range, and happy children who earn their candy money raising, planting, and tending trees and shrubs. Incidentally, the quality of the tea fetched the highest price at the Calcutta auction this past season, while production totals now match previous chemical production figures after a four year transition period.

## Using Off-Farm Nutrient Sources

The key to ecological soil management is recycling nutrients via organic matter. On many farms, mineral nutrients must be added to replace those removed by the sale of crops and livestock. Loss of nutrients in other ways—such as leaching, erosion and fixation in unavailable forms—is minimized when the level of organic matter is high, pH is moderate, and microbial populations are healthy.

When off-farm fertility inputs are needed to supplement nutrients contributed by organic matter management, numerous choices are available. In most cases, materials derived from organic wastes or mined minerals are readily available and desirable. While they may be chosen primarily because they provide particular nutrients observed to be deficient or in high demand by a crop, they generally contain substantial amounts of other major and minor nutrients. When you rely on natural materials to replenish nutrients, there is little risk of creating imbalances in the soil ecosystem or of causing pollution by overapplication of soluble nutrients.

## Purchasing Natural Fertilizer Materials

Calling a product "organic" has suddenly become a hot selling point. There are probably a hundred companies selling 'organic' or 'natural' fertilizers, and every day new ones appear. There is often ample cause for skepticism. Remember that, although legislation exists for labelling food as "organically produced," there is, as yet, no similar protection for consumers of fertilizer products. Some manufacturers think that adding a little fish or seaweed powder makes an otherwise chemically-based fertilizer suddenly "organic." In some cases this is deliberate misrepresentation, but in others it is understandable ignorance.

The more information a manufacturer is willing to provide to you—whether on the label or in response to an inquiry—the more confidence you can have in the integrity of the product. Beware of secret formulas and vaguely described ingredients. Once you know what materials are in the bag, you can make an informed decision about whether you want it in your soil. You may find that cost per pound of nutrients provided is the deciding factor for choosing a particular product from the many options available.

### Fertilizer Or Soil Amendment?

What is the difference between fertilizers and soil amendments? These terms are often used interchangeably, but there is a distinction. When you apply a material primarily to provide nutrients to plants, it is considered a fertilizer. This includes high-analysis organic wastes as well as chemicals of synthetic origin.

Soil amendments, while they might contain the same nutrients as fertilizers, are materials used primarily to improve soil tilth. They may contain mostly organic matter, such as compost, or very slow release minerals, such as greensand. Because their nutrients are not immediately available to plants, or are not considered to be important plant nutrients (e.g. calcium), they may not be sold as fertilizers.

These products can be valuable because of their convenience and uniformity, when you have special soil needs, and when farm-produced compost isn't available in adequate supply. Beware of any "natural" product that offers a double-digit nutrient analysis; very few natural high-analysis fertilizer materials contain over 10% immediately available nitrogen, phosphorus or potassium (see Table 17), and this analysis will be diluted if any of these sources is used in a blended product.

If you are involved in or considering applying for organic certification, you should be careful to check with your certifying agency about the permissability of any ingredient before using it on your soil. Many "natural" products contain materials such as Chilean nitrate or soap wastes to boost nitrate or phosphate analysis, which may be allowed in some programs and not in others. Other natural waste products, such as leather meal, are under consideration. Refer to Chapter 5 for more information about organic certification procedures. If organic certification is not a concern

*Table 17. Some common non-organic sources of mineral fertilizers.*

| Product | Nutrient supplied & percent analysis | Relative cost | Comments |
|---|---|---|---|
| Lime, calcitic | Calcium—100% $CaCO_3$ | Low | Use as directed by soil test recommendations. |
| Lime, dolomitic or Hi-Mag | Calcium & magnesium 90-100% $MgCa(CO_3)_2$ | Moderate-low | Best source of magnesium when lime is needed. |
| Gypsum | Calcium & sulfur 90-100% $CaSO_4 \cdot 2H_2O$ | Low | Used to supply calcium where soil pH is high. |
| Langbeinite | Sulfur: 22% Potassium: 22% (oxide) Magnesium: 18% (oxide) | Moderate-high | Natural mined rock, soluble without posing danger of salt buildup when used according to soil test recommendations. |
| Colloidal phosphate | Phosphorus: 22% $P_2O_5$ Calcium: 27% CaO & trace | Moderate | Analysis varies depending on source. |
| Rock phosphate | Phosphorus, calcium & trace | Moderate | Analysis varies depending on source. |
| Superphosphate | Phosphorus: 20% $P_2O_5$ Sulfur: 14% Some calcium | Moderate | May be less costly than rock powders per unit phosphorus. |
| Wood ash | Potassium: 7% Calcium | Low | Overuse can cause excess potassium buildup and salinity. |
| Granite dust | Potassium: 5% & trace | Low | Very slowly available, but provides long-term reserves. Use mica-rich type only. |
| Greensand (Glauconite) | Potassium: 5% & trace | Moderate-high | Very low availability. |

for you, some of these products can be valuable short-term measures to supplement otherwise inadequate soil nutrient levels.

## Uncomposted Organic Materials

There is a range of uncomposted organic fertilizers available, generally dried and pulverized waste products of food or fiber processing industries. In some cases, livestock feed ingredients such as soybean and alfalfa meal are sold as fertilizers. Many farmers find that using these materials as nitrogen supplements is simpler than putting them through an animal first.

Most of these products are more or less "generic," meaning that dried blood or alfalfa meal packaged by company X is not likely to be different from that sold by company Y. Similarly, mineral fertilizers comprised of finely ground rock powders tend to be comparable from brand to brand, although the quality of some rock mineral deposits is known to vary. Kelp and fish meals may differ depending on species used and how they are processed (e.g., heat and chemical stabilizers). Refer to Table 12 for information about generic organic wastes.

A major consideration in using industrial organic wastes is the possibility of toxic contamination. Dioxin in paper mill sludge, cadmium in leather tankage, and high levels of pesticide residues in cotton gin trash are all of critical concern, in light of the importance of recycling the nutrients in these materials. In some cases, composting the material first will degrade or immobilize the contaminant, but in most cases such materials are not permitted for use in organic production, and many are treated as hazardous wastes under environmental regulations.

Blended products can differ considerably from each other in terms of nutrient content as well as health and environmental criteria. In such cases it is wisest to check out the complete list of ingredients and make sure you feel comfortable about putting all of them in your soil. Then compare the cost; you will generally find that creating your own blend from generic ingredients is much less expensive than paying for the convenience of a specially formulated blended product.

## Compost-Based Products

There is an increasing array of compost-based fertilizer products available on the market. While many are aimed primarily at home gardeners, a few cater to the needs of commercial farmers. Locally produced composts are commonly available in bulk, whether made by municipal yard waste facilities or nearby food processors. Skyrocketing landfill costs in recent years have led many strategically located farmers to set up waste recycling operations to take advantage of free fertilizer materials. In some cases, the tipping fees have ended up generating more income than the farming activities See pages 97–106 for information about farm-scale composting.

Your choice of a commercial compost-based product will depend on its intended use, as well as knowledge of its ingredients, the composting process used, and the reputation of the manufacturer. There is a wide range of organic waste products being composted, and each presents its own challenges for the industrial composting process.

The main headache for large compost producers is achieving uniformity—it's very difficult to get two batches to come out exactly alike in terms of guaranteed analysis. Some companies have worked this problem out by carefully controlling the compost ingredients and process, but most get around it by amending the finished product with uncomposted organic materials or minerals. If uniformity of texture or nutrient content is not a concern, local bulk composts will certainly be more cost-effective.

Composting municipal sludge generally kills diseases that may have been present in the raw sewage, but sludge composts can sometimes contain concentrated amounts of toxic heavy metals—most states forbid sale of contaminated sludge products as fertilizer. Mushroom compost, the spent bedding from commer-

## Federal Law & Acceptable Organic Fertilizers

The use of the word "organic" on a label, according to fertilizer industry guidelines, may simply mean "containing carbon." The term "natural" refers to anything of plant, animal or mineral origin, regardless of what industrial manipulation it has received in the meantime. The Organic Foods Production Act of 1990 (OFPA) gives some guidelines about which kinds of fertilizer materials are and are not acceptable for organic production, but leaves it to the National Organic Standards Board (NOSB) to make recommendations regarding materials to be included on a list of "allowed synthetic" and "prohibited natural" agricultural inputs. This list will go into effect on its publication as a Final Rule, expected in early 1996, but will be subject to revision through a petition process.

While fertilizers derived from natural sources will be generally permitted, the national standards will place certain restrictions on the use of all farm input materials. To be acceptable for organic farming, any commercial fertilizer product will have to fully disclose all ingredients, so a determination can be made that it contains no prohibited materials.

Before the OFPA was passed, many organic certification programs had started to get away from a strict "natural" versus "synthetic" criterion for evaluating fertilizers. Although there is general agreement that man-made materials have a greater potential for harm, it is also recognized that the "ecological profile" of a material is more important than whether it was dug from the ground or is a by-product of an industrial process. The problems associated with certain natural materials have long been acknowledged. It has taken longer for organic advocates to realize that it's not wise to universally prohibit all synthetics. To cite one instance, elemental sulfur, which is used to control plant diseases and to neutralize soil alkalinity, may be naturally mined or derived from industrial stack scrubbers—and telling the difference is very difficult.

For current information about acceptable organic materials, contact the USDA National Organic Program or your organic certifier of choice (see Appendix B).

cial mushroom production, is another potentially residue-laden material. When in doubt, ask for a laboratory analysis showing a clean bill of health for the product.

## Specific Mineral Nutrient Sources

If a farmer needs to correct specific nutrient deficiencies, the most biologically compatible sources of mineral fertilizers are slow-acting rock powders. The advantages of natural rock powders include:

- Less shock to soil organisms than with a heavy dose of soluble compounds.
- Less danger of "burning" plant roots, or of salt buildup in soil.
- Less energy-intensive production methods.
- Unrefined rock deposits contain essential micronutrients.
- Often the cheapest source of the desired nutrient, if total content (including that which is not immediately available) is considered.

Disadvantages of natural rock powders include:

- Large quantities are needed to provide enough available nutrients for immediate crop yields.
- Ineffective in soils that are low in organic matter, and unavailable in alkaline soils.
- Long range planning is necessary for most effective use.
- Unrefined rock deposits may contain undesirable contaminants.
- Considerable ecological costs of mining and transport.

Some major sources of mineral nutrients are, of course, manure and other organic materials. Average analysis tables can be used for rough estimates of what is provided by these sources (see Table 12). Green manures, including weeds, can also contribute needed minerals by converting "unavailable" soil nutrients into organic matter. When calculating fertilizer needs, the usual procedure is to deduct from total needs (as indicated on soil test results) the amount that will be supplied by available manure. The remainder must then be supplied through other sources.

Soluble mineral fertilizers have a place in ecological soil management, especially in the eastern U.S. and other high rainfall areas, where some leaching loss is unavoidable. When selecting such materials, choose the alternative least likely to harm soil organisms. Therefore avoid highly concentrated soluble salts, whether natural or synthetic. When adjusting nutrient levels by applying fertilizers, be aware of *balance.* It is better to uniformly underfertilize than to oversupply one nutrient in relation to the others. Theories about nutrient balances are explained in Chapter 2.

Reliance on soil test recommendations is no substitute for a thriving, diverse soil ecosystem. Minerals that promote this goal by satisfying soil organisms should always be chosen in preference to those that provide nutrients needed by plants, but which may inhibit biological activity in the soil. Moreover, as pointed out earlier, most recommendations are only approximations of soil needs. *Building humus is the only way to ensure healthy crops under virtually all circumstances.*

## Major Cation Nutrients: Calcium, Magnesium & Potassium

**Calcium,** a building block of plant tissue, is also essential for the health of microbes. University soil labs rarely test for calcium, but it is generally supplied through lime applications. In most cases it is safe to assume that soil calcium is low if you have a low pH. Lime recommendations are usually determined by a combination of cation exchange capacity, pH, and in some cases, aluminum levels. Standard liming recommendations can be followed without hesitation—as long as you watch magnesium levels as well.

Some sources of calcium include calcitic lime, dolomitic lime (also contains magnesium), colloidal phosphate, bone meal, gypsum (mostly used where pH is high, but also useful with neutral and lower pH), and wood ashes (in small quantities only). Calcium salts, such as chlorides, and hydrated lime or other strong bases, are undesirable as sources of calcium.

**Magnesium** was largely ignored in conventional fertility recommendations until recently. It behaves similarly to calcium in the soil, and is just as deficient in highly leached, acid soils.

Since magnesium occurs most commonly in combination with calcium or potassium in natural rock deposits (dolomite and langbeinite), it can be difficult to correct a magnesium deficiency if your soil is already well-supplied with calcium and potassium. In that case the best approach may be foliar application of a chelated magnesium source to avoid plant deficiencies. Magnesium sulfate (epsom salts) is sometimes used, but this is only economical on small areas and must be done with great care.

Some sources of magnesium include dolomitic lime and langbeinite (Sul-Po-Mag™ and K-Mag™ are trade names).

**Potassium** is present in most soils, but is often unavailable to plants. It is also one of the few nutrients that is not an inherent component of organic molecules, though some organic materials contain substantial amounts. Organic matter improves potassium availability mainly by raising the soil's cation exchange capacity (CEC). The only way to store a significant amount of potassium in soil reserves available to plants is on colloidal exchange sites.

Available soil potassium not held by the CEC is quite soluble. Any excess applied as fertilizer will either be lost through leaching or will cause "luxury" consumption by plants: they will keep absorbing it beyond their needs before it is adsorbed onto colloidal exchange sites. This is why some soil labs will recommend "split applications" and side dressing for potassium.

Large enough quantities of granite dust for use as a potassium source can sometimes be obtained. (Only that which is rich in mica, as opposed to feldspar, is useful as a fertilizer.) Because of its fine particle size, the mineral weathers relatively quickly, releasing about 2% of its weight as available potassium during the course of a growing season.

Higher analysis sources, which are quite soluble, are more feasible for correcting large potassium deficits. The ecological acceptability of a potassium fertilizer depends largely on the other elements in the compound. The most common—and the cheapest—potassium carrier is muriate of potash (potassium chloride). Although it is a natural substance, it is undesirable due to its desiccating and salinizing effects, and residual chloride ions.

Some sources of potassium include Sul-Po-Mag™ or K-Mag™ (langbeinite), potassium sulfate (sometimes available in a natural form), wood ash, granite dust, and kelp meal.

## Major Anion Nutrients: Nitrogen, Phosphorus & Sulfur

Although there is such a thing as "anion exchange capacity," it is of little significance for fertility management. Mineralization of organic matter is the major source of anion nutrients, other than soluble forms applied directly to crops.

What follows are guidelines for using supplemental sources of these nutrients when soil organic matter reserves are low. This is only acceptable as a stopgap, short-term strategy: the first priority is always on supplying crop needs through organic matter and humus.

**Nitrogen** requirements were mentioned earlier under organic matter management. As noted, additional sources of nitrogen may be helpful when mineralization of organic matter cannot match crop needs (such as under cold, wet conditions or during drought), or on transitional soils where humus has not yet reached desired levels.

Higher crop yields may be obtained by using more nitrogen than is naturally provided by organic matter. However, such gains may be produced at the cost of poorer crop quality in terms of percent dry weight, insect resistance, and storage qualities.

There may also be times when crop production is imperiled by nitrogen deficiency, especially on fields with low levels of organic matter. An immediate dose of soluble nitrogen can be provided in several ways. The costs and benefits of each must be considered in light of the situation that confronts you.

Crops can be side-dressed with any available organic material having a heavy nitrogen content (see Table 12), or they can be given a heavy application of manure or compost (twenty tons or more per acre). There are also commercial "organic" formulations that use blood meal, fish emulsion, leather meal, or a similar highly nitrogenous waste product to provide supplemental nitrogen. However, these are usually more costly than manure and may contain harmful contaminants such as heavy metals. Farmers not concerned with organic certification may also consider using certain soluble nitrogen sources, such as sodium nitrate (also known as Chilean). If you are considering entering the organic market, you should refer to the discussion beginning on page 122 before deciding to apply any questionable fertilizer material.

In the long run, the goal of your management program should be to eliminate the need for any such fast-acting "rescue chemistry."

**Phosphorus:** The problem of maintaining an adequate supply of phosphate is a tricky one. Debates over the best phosphate sources, and how best to ensure its availability, are ongoing in both ecological and conventional circles.

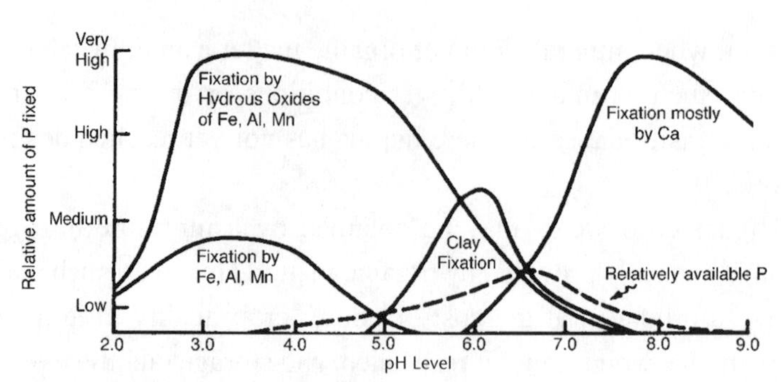


*Figure 18. Phosphorus availability*

Most soil phosphorus reserves are present in organic matter, so adequate supplies and good mineralization are, as usual, the best insurance for fertility. Organic matter also improves the availability of mineral phosphate sources, whether these are rock powders or artificially concentrated forms. Other factors also influence mineral phosphate availability:

- **Soil Reaction**—Phosphorus from any source quickly becomes insoluble at pH extremes. The optimal pH range for phosphorus availability is between 6.0 and 7.0, but rock phosphate is more available at a pH below 6.0 (see Fig. 18).
- **Nitrogen**—Increasing available nitrogen improves availability of existing phosphates, up to a point.
- **Mycorryhizal Fungi**—A healthy soil ecosystem promotes the growth of these symbiotic organisms that make phosphorus reserves more available to plant roots.

On soils where phosphate needs are moderate, an application of one ton of hard rock or colloidal phosphate per acre should satisfy crop phosphate requirements for four to five years. Choice of hard rock or black mined phosphate versus colloidal phosphate depends mainly on availability, price and convenience.

Acid soils are more conducive than alkaline soils to utilization of rock phosphate. For maximum effectiveness, spread rock phosphate uniformly and incorporate it to ensure its availability to soil microbes. The finer the particle size, the more readily available it will be. Rock phosphate is most available when applied:

- With manure, before spreading (also conserving nitrogen), or in compost.

- Before incorporating sod or green manure.
- Prior to seeding legumes in rotation—especially clovers and alfalfa, which are able to make good use of phosphate in this form.

When soil phosphate reserves are low and pH is high, non-organic farmers may want to use artificially concentrated phosphates, such as ordinary super (which is becoming harder to obtain), especially for healthy legume stands. Soluble phosphates do not remain soluble for long, before they are complexed by soil minerals or leached. No matter in what form phosphate is applied, microbial activity is essential for phosphate release and crop growth. Triple superphosphate should be avoided because of its concentrated, acidified nature.

**Sulfur,** like nitrogen, is an essential part of protein. It can be deficient in soils with depleted humus. Unlike nitrates, sulfates in soluble mineral form pose no threat to soil organisms.

Significant amounts are often found in rainwater, especially in areas that suffer from acid rain. Sulfur deficiencies can also be created when farmers use high analysis phosphate and nitrate fertilizers (triple super, ammonium nitrate and urea), instead of lower grade sources (ordinary super and ammoniated superphosphate) that supply sulfur as well.

If organic matter levels or mineralization rates do not supply enough nitrogen, a supplemental sulfur source may also be needed. A good source of organic nitrogen, however, will generally supply sulfur needs as well. Otherwise, any ecologically compatible sulfur carrier, such as gypsum, can be used, depending on the requirements for any other nutrients it may contain.

In general there is little risk of oversupplying sulfur in relation to nitrogen or carbon, so nutrient sources that also contain sulfur can be applied without fear of burning plants or harming microbial populations. Because it forms sulfuric acid in the soil solution, sulfur also has the effect of lowering soil pH.

## Micronutrient Fertilizers

The behavior and function of micronutrients in soil and plants is complex and not well understood. This alone is a good reason to rely on organic matter and soil biological activity (supplemented with foliar fish and seaweed) to supply plants with properly balanced amounts of micronutrients.

However there may be instances when you have cause to suspect a specific micronutrient deficiency—perhaps plant symptoms, previous crop problems or soil test results. The safest way to confirm such a suspicion is with plant tissue analysis. Most soil tests for micronutrients (boron excepted) are less accurate than plant tissue analysis.

Deficiencies diagnosed in growing crops or orchards are most effectively treated with foliar fertilizers. Deficiencies diagnosed before planting can be corrected by applying specific micronutrient fertilizers, usually available as chelates or frits, to the soil.

**Chelates** are organic molecules, such as lignin sulfonates, complexed with cation micronutrients to mimic the action of humus. The nutrients are readily available to plants, but don't react with other elements or leach out.

**Frits** are micronutrient salts (anion or cation) which have been fused with glass and shattered into a powder. The nutrients are released slowly as the glass dissolves, becoming available to plants at a gradual rate.

Micronutrients must be evenly applied in correct amounts to prevent toxicity. The best time to apply them is while incorporating fresh organic matter, or before seeding a green manure crop. See Table 18 for details on some micronutrient fertilizers.

## Foliar Fertilization

"Feed the soil, not the plant" is a principle of organic farming, but there are circumstances that warrant direct plant feeding. All plant leaves absorb nutrients such as carbon and sulfur in gas-

*Table 18. Solid micronutrient fertilizers.*

| Nutrient | Crops with high demand | Sources and analysis | Maximum application |
|---|---|---|---|
| Boron (B) | Apples, brassicae, celery, beets, clover, & alfalfa | Borate 65: 20% B<br>Borax: 11.3% B<br>Boron frits: 2-11% B | 2.5-5.0 lbs/acre B for high demand crops. |
| Molybdenum (Mo) | Legumes, cauliflower, lettuce, spinach | Sodium molybdate: 39% Mo<br>Ammonium molybdate: 54% Mo<br>Molybdenum trioxide: 66% Mo | 0.5-1.0 lbs/acre Mo. |
| Manganese (Mn) | Legumes, small grains, potatoes | Manganous oxide (MnO): 41-68% Mn<br>Manganese sulfate: 26-28% Mn | 7-15 lbs/acre Mn.<br>Foliar application often preferred. |
| Copper (Cu) | Onions, beets, carrots, corn, oats, fruit trees | Copper sulfate: 25.5% Cu, 12.8% S<br>Copper chelates: 9-13% Cu | 3-6 lbs/acre Cu in sulfate form.<br>0.8-2.4 lbs/acre Cu in chelate form. |
| Zinc (Zn) | Corn, fruit & nut trees, legumes | Zinc sulfate: 35% Zn<br>Zinc oxide: 78% Zn<br>Zinc chelates: 9-14% Zn | 20-40 lbs/acre Zn. |

eous form. Leaves are also able, in some situations, to absorb dissolved minerals and translocate them to sites where they are needed in the plant. As we have seen, the use of certain soluble mineral fertilizers may sometimes be warranted as a short-term measure for ensuring healthy crops within a long-term soil building program.

Foliar fertilizers may be used as a stopgap remedy for definite nutrient deficiencies until the soil is able to supply them. Foliar supplementation may also be necessary as an annual practice in some soil and climate situations. In cold northern soils, foliar fertilizers may be necessary each spring to supply nitrogen and phosphorus. In alkaline and high organic matter (peat and muck) soils, micronutrient sprays like iron may be necessary since certain soil minerals may never be available under these conditions.

Liquid seaweed extract and fish emulsion are popular among ecological farmers and gardeners. The seaweed or kelp plant itself illustrates the principle of foliar feeding. It is anchored to the seabed by holdfasts, not roots, and feeds through its leaves from mineral-rich ocean water. Since it balances these mineral inputs and binds them into organic molecular structures (chelation), the cell sap extract becomes a very useful foliar material.

There are many factors to consider when you manipulate plant growth through foliar fertilization. Some minerals (such as calcium and boron) are easily absorbed but poorly translocated, while others (such as iron and copper) move easily through the vascular system but don't readily cross the leaf epidermis (see figure 19). Temperature, light, oxygen, pH, and energy availability all affect leaf absorption, but the most important factor is leaf surface moisture.

The best time for foliar fertilization is in the early morning of a cloudy, humid day. Effectiveness increases with the fineness of the spray droplets. A small amount of surfactant added to the mix will deionize the leaf surface, improving absorption of the mate-

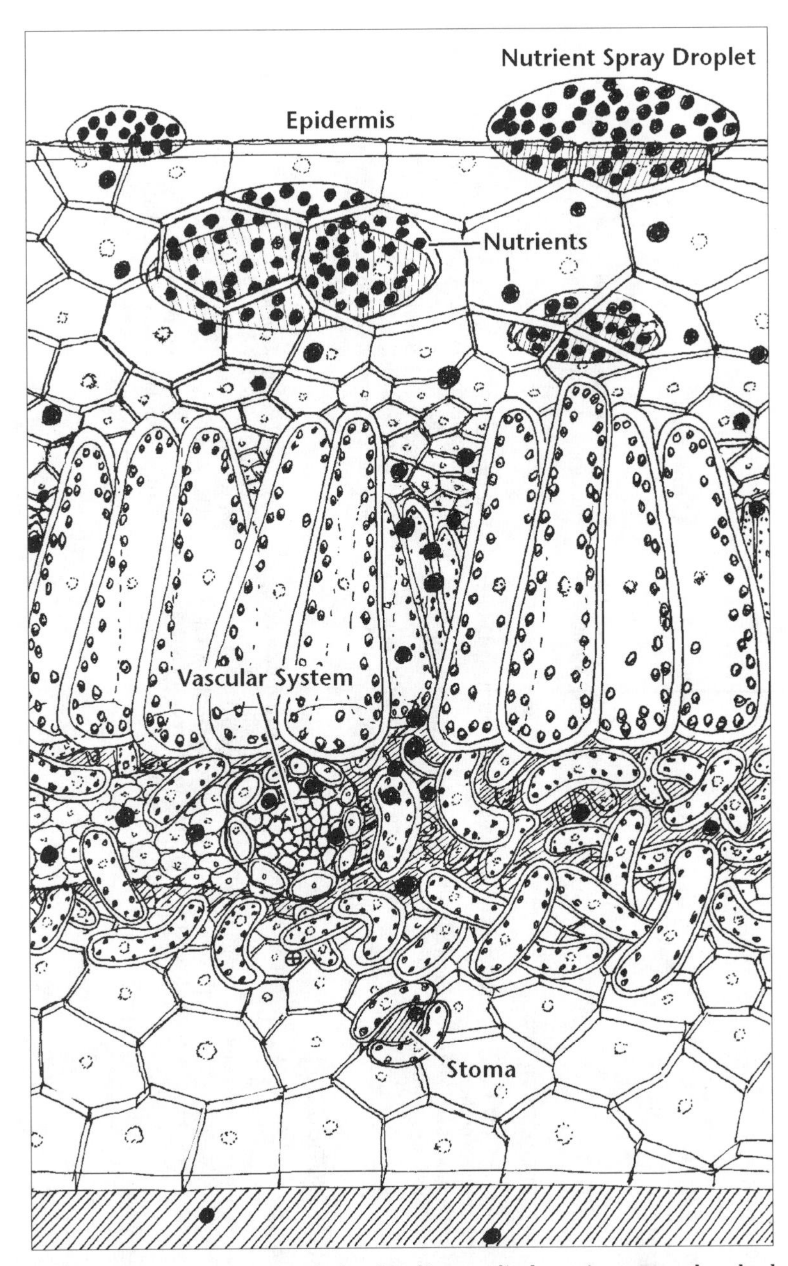


*Figure 19. Cross-section of a leaf. Foliar applied nutrients are absorbed through the epidermis and stoma and translocated through the vascular system. (Drawing by Stewart Hoyt.)*

*Table 19. Foliar fertilization.*

| Product & Analysis | When to use | Comments |
|---|---|---|
| FISH EMULSION | | |
| 4-4-1 (N-P-K) varies by manufacturer | Transplanting, and any time crops need an extra N & P boost, as in early spring: Dilute 20:1. | As a general rule, one gallon will treat one acre of crops at a time (2-4 gal/season). |
| LIQUID SEAWEED CONCENTRATE | | |
| Contains more than 1% of: K, Ca, Na, Cl, S, and about 50 other minerals in trace amounts. Also plant hormones such as auxins & cytokinins. | Transplanting & rooting cuttings: Dilute 25:1.<br>Flower bud & petal fall (fruit crops): Dilute 40:1.<br>To alleviate stress (from drought, cold, etc.) and promote strong regrowth of hay: Dilute 40:1. | Use at most 1.5 gallons per acre per season.<br>If blending fish & seaweed together, dilute one before adding the other.<br>It is sometimes helpful to use a surfactant (wetting agent) to aid even spray distribution.<br>Best time to spray any foliar nutrient is early morning or evening, never in full sun. |
| SOLUBLE MICRONUTRIENTS | | |
| All sulfates & chelates of: Mn, Fe, Cu, & Zn can be used in foliar applications, as well as Mo salts. B is available as Solubor™ (20.5% B) or boric acid (17.5% B). | All should be used only when indicated by plant tissue analysis, or when a soil deficiency is known to exist.<br>In general, a lower rate of application (in lbs/acre) is needed to correct deficiencies than when soil applied. | B is rarely recommended for foliar application, except on apples.<br>Mo solutions can be used to soak legume seeds when inoculating.<br>Always explore the possibility that deficiencies are due to soil imbalances. |

rial. Timing of sprays should also coincide with with the stage of plant growth when nutrients are in high demand by the plant but root intake is low.

Leaves can only accept small amounts of mineral materials. Therefore micronutrients lend themselves to foliar application, but significant amounts of major nutrients cannot be supplied in this way. When some elements (such as Ca, Mg, Fe, Mn, Mo, Cu) are chelated, they are more easily absorbed than as a mineral salt. There is a controversy in organic circles about "synthetic" chelates, and some research indicates that these chelates may tie up plant minerals after delivering their nutrient. The most acceptable synthetic chelating agent for ecological farmers is lignin sulfonate. Natural chelates, like liquid seaweed extract or mineral salts chelated by processing with molasses, are generally accepted.

Another controversial aspect of foliar spraying is the use of growth-promoting substances. Liquid seaweed has been touted for years by ecological agriculturists as a general purpose trace mineral spray and growth promoter. It contains several hormones, including auxins, giberillins and cytokinins. All of these are chemical messengers produced naturally in minute quantities by the plant. Cytokinins are of particular importance, since they increase the speed of cell division. Introducing hormones from elsewhere increases growth rates or breaks dormancy earlier than the plant ordinarily would do.

# 5

# The Marketplace & Organic Certification

There are many reasons for the increasing interest in ecological management systems, high among them the marketplace demand—and price premium—for certified organic foods. In producing for this market, certain standards must be met. These can vary somewhat according to the certifying organization, but in Europe, Argentina, Canada, or the U.S., standards must now conform in their minimum criteria to government regulation. In the international arena the Codex Alimentarius Commission of the United Nations has been preparing guidelines since 1990 that may set the protocol for international trade in organic food. In the U.S., the passage of the Organic Foods Production Act (OFPA) in the 1990 farm bill set the rules, with detailed regulations to be enacted.

While many provisions of this legislation are controversial, it has for the first time established a unified set of national standards which precisely define organic production methods. It is also significant in placing heavy emphasis on soil quality and a sound soil management plan as criteria for evaluating applicants for certification. The final regulations will include stipulations about one of the more contentious aspects of soil management: Which materials (including many kinds of farm inputs such as pest controls and livestock feed, in addition to fertilizers) are acceptable for use in organic production, and on what basis should new ones be considered? In order to understand this issue, some background on the development of organic certification programs is helpful.

The organic market arose from the traditions of organic agriculture, which chose the term "organic" to emphasize the importance of living systems and their natural growth processes, with a focus on building soil organic matter. In the modern era, harmful pesticides and synthetic fertilizers have been used with disregard for the complex biological needs of living soil, resulting in many health and environmental problems. In reaction to this, organic farming advocates maintain that only naturally occurring fertilizer materials and pest controls are acceptable, as opposed to "chemical" or "synthetic" ones.

As ecological thinking has become more sophisticated, many agricultural leaders recognize that creating dichotomies such as "organic" versus "chemical" is inaccurate and a hindrance to more widespread adoption of this approach. Because agriculture inevitably manipulates the environment, there is no place to draw the line of "natural," short of eliminating agriculture altogether and becoming hunter-gatherers.

Several organic certification programs have, in fact, started to get away from a strict "natural" versus "synthetic" criterion for evaluating fertilizers. Although there is general agreement that man-made materials have greater potential for harm, organic farmers have come to recognize that the "ecological profile" of a material is more important than whether it was dug from the ground or is a by-product of an industrial process. The problems associated with certain natural materials have long been acknowledged. For example most certification programs have consistently prohibited the application of raw manure directly before planting, and the use of mined muriate of potash due to its high salt index and chlorine content.

It has taken longer for organic advocates to realize that it's not wise to universally prohibit all synthetics, regardless of ecological and health considerations. While all synthetic pesticides are prohibited because of their toxicity, mutagenicity, and persistence in the food chain (and even here there are cases where synthetics, such as mating disruptive pheromones, are less damaging to soil life and natural predator populations than the accepted natural alternatives), certain synthetically derived fertilizer products may be the most ecologically sound approach to correcting a soil problem. Superphosphate and urea, for example, are two materials which some argue should be an option in an ecological fertility program, under carefully specified conditions. As our friend Eliot Coleman says, "Nothing is as stifling to success in agriculture as inflexible adherence to someone else's rules."

Although the old distinction of natural versus synthetic is included in the OFPA, it acknowledges the justification of these arguments by mandating a process for evaluating materials which, though natural, should not be accepted for organic farming, as well as those which may be permitted although they are synthetically derived. In doing this, the national advisory board is relying on a complex theoretical format which basically asks the following questions of new materials:

- What effects will it have on plant growth and health?
- Is it safe for farmworkers to handle?
- Will it affect food safety or quality?
- What will it do to (or for) soil organisms?
- Can it lead to air or water pollution?
- Does its mining, manufacture, or transport harm the environment?
- How energy-intensive is it to make?
- Does it help recycle an existing organic waste product?
- Is there a feasible alternative which may be more environmentally benign?

The most important service rendered by this model is to tell us that answers simply don't exist for many of these questions, even for some of the most commonly used fertilizers. This is because until now scientists and fertilizer manufacturers have tended to focus their research on one question: Does it increase yields? Until research exists to uphold the argument for use of some synthetics, there may be discrepancies between what is "certifiably organic" in the marketplace and what is "ecologically appropriate" on the farm.

If you are planning to sell to the organic market, there should be little difficulty in using any of the methods outlined in this book. It is essential, however, that you contact the certification program you intend to work with (information about how to find them is given in Appendix B) about specific requirements and regulations before you use any material you are unsure about. OFPA currently requires a transition period of three years after use of any prohibited material (including fertilizers) until a given field can be approved for organic certification.

Whatever the requirements of your certifying organization, proper recordkeeping and a farm plan are musts (see pages 53–56). It must be possible to trace, on paper, what fertilizers and pest control materials were purchased, where and when they were applied, and at what rates. In addition, you should have some system for keying sold crops to their field of origin by lot number. Inventories of purchased materials on hand, especially if they fall into regulated or prohibited categories must be scrupulously recorded. If some portion of your farm is not under organic certification, there must be no chance of intermingling products which are organic with those which are not. Taking great care with paperwork or computerized recordkeeping avoids possible doubt or misunderstanding and creates a valuable resource for your own planning and management.

Even federal legislation and its regulatory interpretation is not written in stone. The rules should change as better information becomes available. It is, however, important to develop and maintain marketplace credibility for organic foods, and therefore unwise to risk confusing consumers with contradictory definitions. The process of marketplace education goes hand-in-hand with the education of producers, who must balance their need for ecologically sound, profitable management tools with their responsibility to uphold consumer trust in their products. The best guarantee for any consumer is still a face-to-face, personal relationship with the farmer. The wisest strategy for the farming community is an alliance with ecologically aware consumer groups.

Participation in organic certification programs will increase your planning, marketing and communication skills. The organic marketplace offers an opportunity for ecological farmers to improve their profitability. Widespread adoption of ecological agriculture can only be assured by making it profitable.

# Appendix A

## Glossary

**Adsorption:** Attraction of charged particles to a solid surface of the opposite charge, as in the attachment of cations to soil colloids.

**Aerobic:** A process that requires the presence of free oxygen, or a condition in which free oxygen is present.

**Aggregate:** A collection of soil particles which holds together, forming the basis for good soil structure.

**Allelopathy:** Suppression of one plant species by another through secretion of phytotoxic exudates.

**Anaerobic:** A process that does not require free oxygen, or a condition in which free oxygen is excluded.

**Anion:** A negatively charged chemical element or radical which is acid-forming.

**Autotroph:** An organism, usually a green plant, that uses inorganic materials to synthesize it sown food through the process of photosynthesis.

**Base Saturation:** The proportion of total cation exchange sites in a given soil occupied by bases (cations other than hydrogen or aluminum).

**Bio-Dynamics:** A school of agricultural thought based on the spiritual teachings of Rudolph Steiner, that views the living interrelationships of soil, farm and community as an organic whole.

**Biomass**: The total weight of vegetation growing on a given area of soil.

**Bioregion:** A geographic area defined by common ecological characteristics, such as climate, soil types, topography, predominant vegetation, and water system.

**Brix scale:** A measure of percent of sugar content of a liquid, usually cell sap or fruit juice.

**Calibrate:** To establish the relationship between laboratory soil test values and yield response from incremental rates of the nutrient applied in the field.

**Capillary action:** The ability of soil to wick up moisture from deeper levels as it is needed, made possible by good structure.

**Carbon:nitrogen ratio:** The proportion of carbon to nitrogen by weight in any organic matter. The optimum level for biological activity in raw organic matter is between 20-30 to 1.

**Cation:** A positively charged chemical element or radical, usually alkaline-forming.

**Cation exchange capacity (CEC):** The capacity of the colloidal particles in a given soil to hold positively charged ions (cations).

**Chelate**: A compound consisting of a metallic element bound within a complex organic molecule, or the process of forming such a compound.

**Climax ecosystem:** The plant/animal/soil community that is most stable and well-adapted to a particular geographic locale.

**Colloid:** A mass of very fine particles having a high ratio of surface are to weight, and characterized by a gooey consistency. Examples include protoplasm, clay, humus and mayonnaise.

**Compost:** A process that uses any one of several methods to speed up the decomposition of raw organic matter, usually by piling, aerating, and moistening. Also, the crumbly, nutrient-rich product of this process.

**Consumers:** Organisms that obtain nourishment by consuming the products or remains of other living organisms.

**Correlate:** To express the relationship between amounts of nutrient extracted by laboratory tests and uptake by plants in the greenhouse or field.

**Cover crop:** A crop used to protect soil from erosion, at the same time building organic matter and controlling weeds.

**Decomposers:** Organisms, usually soil bacteria, that derive nourishment by breaking down the remains or wastes of other living organisms into simple organic compounds.

**Ecology:** The science that studies the interrelationships between living organisms and their environment.

**Exudates:** Biochemical compounds secreted by plant roots.

**Fertilizer:** Any material added to the soil for the purpose of providing essential nutrients to plants.

**Fixation:** The binding of a nutrient into a more stable form that may either be less available to plants (as when phosphorus is fixed by calcium and aluminum), or more available to plants (as when atmospheric nitrogen is biologically fixed by bacteria).

**Green manure:** A crop that is incorporated into the soil before reaching full maturity, for the purpose of soil improvement.

**Hardpan:** A heavily compacted subsoil layer, sometimes naturally occurring, but usually created by poor soil management practices.

**Heterotroph:** An organism whose nourishment depends on organic materials produced by other living organisms.

**Horizon:** A distinct layer of soil (parallel to the surface) that characterizes the soil-forming process in a particular locale.

**Humus:** The fragrant, spongey, nutrient-rich material resulting from decomposition of organic matter.

**Immobilization:** The conversion of an element to a form that is not available for biological processes or chemical reactions. Soil nutrients may be immobilized in microbial or plant tissues, or in very stable chemical complexes.

**Inoculant:** The spores of the desired strain of *Rhizobia* bacteria, applied in powder form to the appropriate legume seed when planted. Also, any material of high microbial content added to soil or compost to stimulate biological activity.

**Leaching:** The downward movement through soil of chemical substances dissolved in water.

**Legume:** A member of the plant family that includes clover, alfalfa, beans and peas, whose roots host nitrogen-fixing *Rhizobia* bacteria in a symbiotic relationship.

**Macronutrient:** A plant nutrient needed in substantial quantities, including carbon, nitrogen, phosphorus, sulfur, calcium, magnesium and potassium.

**Metabolism:** The sum of the biochemical processes of growth, maintenance, and energy transformation carried out by a living organism.

**Micronutrient:** A plant nutrient needed in very small quantities, including copper, zince, iron, manganese, boron and molybdenum.

**Mineralization:** The release of soluble minerals and simple organic compounds through the decomposition of organic matter.

**Mucigel:** A gelatinous material surrounding the root, created by the root-soil-microbe complex.

**Mycorrhizal association:** A symbiotic relationship between mycorrhizal fungi and plant roots, in which soil phosphorus is made more available to plants.

**Nitrogen cycle:** The process by which nitrogen passes from a gaseous state, through living tissue in various organisms, and back into the atmosphere. Carbon and sulfur undergo similar cycles.

**Organic matter:** The remains, residues or waste products of any living organism.

**Parent material:** The raw material—native rock or organic matter—from which topsoil is created through biogeochemical processes and weathering.

**Pathogen:** An organism capable of causing disease in other organisms.

**pH:** The concentration of hydrogen ions in a solution that determines its level of acidity or alkalinity. A pH of 7.0 is neutral; lower numbers indicate acid, and higher numbers indicate alkaline conditions.

**Phytotoxin:** A substance toxic to plants.

**Plowpan:** A heavily compacted subsoil layer created by overuse of a moldboard plow.

**Profile:** The pattern of soil horizons, from topsoil to subsoil, that identifies a particular soil type.

**Pore space:** The space between soil particles where air and water circulate.

**Producers:** Organisms, generally green plants, that can make their own food from inorganic materials, usually through the use of radiant energy in photosynthesis.

**Raw manure:** Manure that has not yet decomposed, containing highly soluble nitrogen and potassium.

**Refractometer:** A hand-held instrument that measures the sugar content of any liquid.

**Rhizobia:** Nitrogen-fixing bacteria that live in symbiosis with legumes.

**Rhizosphere:** The area immediately surrounding plant roots, where the highest level of soil biological activity exists.

**Soil amendment:** Any material added to the soil in order to enhance soil biological activity.

**Structure:** The physical arrangement—shape, size, and stability—of soil particles.

**Symbiosis:** A mutually beneficial relationship between two living organisms, such as plant roots and rhizobia.

**Synergy:** The mutual combined enhancement of two or more factors (such as soil nutrients), which exceeds the expected sum of individual effects.

**Texture:** The proportions of sand, silt and clay in a particular soil.

**Tilth:** The physical quality or condition of soil, similar to the health of a living organism.

**Valence:** The electrical charge of a chemical ion, indicating the number of electrons it will accept (for anions) or give up (for cations) in a chemical reaction.

**Volatilization:** The escape of chemical elements into the atmosphere after being transformed into a gaseous state.

# Appendix B

## Organizations & Periodicals

The number of organizations and publications serving ecological agriculture is proliferating rapidly. A few key contact groups are listed here, but more extensive and local listings may be found in the following directories, or by contacting one of the national organizations.

### Directories

***Healthy Harvest: A Global Directory of Sustainable Agriculture & Horticulture Organizations***
agAccess
P.O. Box 2008
Davis, CA 95617
(916) 756-7177 FAX: (916) 756-7188

***National Organic Directory: A Guide to Organic Information & Resources***
Community Alliance with Family Farmers
P.O. Box 464
Davis, CA 95617
(916) 756-8518 FAX (916) 756-7857

***Catalogue of Healthy Food***
Edited by John Tepper Marlin with Domenick M. Bertelli
Bantam Publishing, 1990

***The Organic Resource Guide***
Canadian Organic Growers
Box 116, Collingwood
Ontario, Canada L9Y 3Z4
(705) 444-0923

### National & International Organizations & Periodicals

***Acres, USA: A Voice for Eco-Agriculture***
Box 9547
Raytown, MO 64133
(816) 737-0064
Monthly publication, with books and annual conference.

**Alternative Farming Systems Information Center**
National Agriculture Library, Room 111
10301 Baltimore Boulevard
Beltsville, MD 20705
(301) 344-3704
Bibliographies & literature searches on all aspects of alternative agriculture.

**Appropriate Technology Transfer for Rural Areas (ATTRA)**
P.O. Box 3657
Fayetteville, AR 72702
(800) 346-9140
(501) 442-9824 FAX: (501) 442-9842
Information service funded through the National Center for Appropriate Technology (NCAT). Information packets on many aspects of sustainable agriculture.

**Bio-Dynamic Farming & Gardening Association**
P.O. Box 550
Kimberton, PA 19442
(215) 935-7797 FAX (215) 983-3196
Membership organization with chapters throughout North America, quarterly publication, books, films.

***Biological Agriculture & Horticulture***
P.O. Box 97
Berkhamsted, Herts
HP4 2PX, England
Academic journal featuring the most advanced European work.

**Ecological Agriculture Project**
Box 225
Macdonald College
Ste. Anne-de-Bellevue, Quebec
H9X 1CO, Canada
(514) 398-7771 FAX (514) 398-7990
Bibliographies, article files, publications, audiovisuals.

**International Federation of Organic Agriculture Movements (IFOAM)**
General Secretariat
c/o Okozentrum Insbach
D-66636 Tholey-Theley, Germany
49 (0) 6853-5190 FAX 49 (0) 6853-3011
Membership organization with conferences, newsletter and publications.

***Journal of Soil & Water Conservation***
7515 N.E. Ankeny Rd.
Ankeny, IA 50021
(515) 289-2331 FAX (515) 289-1227

**Organic Farming Research Foundation**
P.O. Box 440
Santa Cruz, CA 95061
(408) 426-6606 FAX (408) 426-6670
Nonprofit foundation dedicated to fostering the improvement & widespread adoption of organic farming practices. Contact them for grant guidelines & applications.

**Organic Trade Association**
P.O. Box 1078
Greenfield, MA 01301
(413) 774-7511 FAX (413) 774-6432
Trade association for the organic food industry, offering political action, consumer information, newsletter, and support for organic marketing.

**Rodale Institute**
22 Main Street
Emmaus, PA 18098
(610) 967-8108 FAX (610) 967-8959
Nonprofit educational arm of Rodale Press.

**Wallace Institute for Alternative Agriculture**
900 Edmonston Road, Suite 117
Greenbelt, MD 20770
(301) 441-8777
National membership organization with newsletter, conferences, and scientific journal for alternative agriculture.

**Woods End Agricultural Institute**
Box 4050
Old Rome, Mount Vernon, ME 04352
(207) 293-2457
Research, publications, soil testing, and consultation.

## Organic Certifiers

For a complete listing of private and state organic certification organizations and the status of implementation of the National Organic Program, contact:

**National Organic Program**
USDA/AMS/TMD
Room 2510-South, P.O. Box 96456
Washington, D.C. 20090
(202) 720-3252 FAX (202) 205-7808

## Organic Farm Organizations

The following regional organic farm organizations offer publications, conferences, workshops, and other education, in addition to organic certification services. Consult one of the directories at the beginning of this section for a complete listing.

**Carolina Farm Stewardship Association**
115 W. Main St.
Carrboro, NC 27510
(919) 968-1030

**Community Alliance with Family Farmers (CAFF)**
P.O.Box 464
Davis, CA 95617
(916) 756-8518 FAX (916) 756-7857

**Northeast Organic Farming Association (NOFA)**
411 Sheldon Road
Barre, MA 01005
Phone: (508) 355-2853
Chapters in CT, NH, NJ, NY, RI, VT.

**Northern Plains Sustainable Agriculture Society**
Box 36
Maida, ND 58255
(701) 256-2424

**Ohio Ecological Food & Farming Association**
65 Plymouth St.
Plymouth, OH 44865
(419) 687-7665

**Oregon Tilth**
31615 Fern Rd.
Philomath, OR 97370
(503) 929-6742 FAX (503) 929-6743

# Appendix C

## Bibliography

Albrecht, W.A. *The Albrecht Papers I & II.* Acres USA, Raytown, MO, 1975.
These books are collections of Albrecht's viewpoints on soil chemistry, soil fertility, and plant, animal and human nutrition.

Ardapple-Kindberg, Eric & Beth. *Soil Fertility for Organic Farmers.* Keyline Environmental Management Systems, Bass, AR, 1991.
A short guide to practical farming.

Balfour, E. *The Living Soil and the Haughley Experiment.* Faber & Faber, London, 1975.
This edition of the 1948 classic summarizes the 32 year organic versus chemical trials at Haughley.

Berry, W. *The Unsettling of America: Culture and Agriculture.* Sierra Club Books., San Francisco, 1977.
This Kentucky farmer/poet has been a major spokesperson for a sane approach to agriculture. His poetic eloquence is grounded in practical knowledge.

Coleman, Eliot. *The New Organic Grower: A Master's Manual of Tools & Techniques for the Home & Market Gardener.* Chelsea Green Publishers, 1989.
Good basic manual drawn from Coleman's 25 years of experience.

Darwin, C. *The Formation of Vegetable Mould Through Action of Worms, with Observations on Their Habits.* John Wiley & Sons, New York, 1978.
This book, which Darwin felt to be his most important work, is still an unsurpassed account of the earthworm's role in the creation of soil fertility.

Faulkner, E. *Plowman's Folly.* University of Oklahoma Press. 1943.
An impassioned plea to end the widespread use of the moldboard plow.

Fryer, Lee. *The New Organic Manifesto.* Earth Foods Associates, Wheaton, MD, 1986.
——. *Food Power From the Sea.* Mason & Charter, New York. 1977.
Lee Fryer, a veteran of the fertilizer industry, provides an approach for organic growing and marketing.

Fukuoka, Masanobu. *The Natural Way of Farming.* Japan Publications, Inc. NY, NY, 1987.
——. *The One Straw Revolution.* Rodale Press, Emmaus, PA. 1978.
Fukuoka's techniques mimic nature in the design and management of farming systems.

Gershuny, Grace. *Start with the Soil.* Rodale Press, 1993.
Straightforward guide to organic soil improvement.

Granatstein, David. *Amber Waves.* Washington State University, 1992.
A sourcebook for sustainable dryland farming.

Howard, A. *An Agricultural Testament.* Oxford University Press, London. 1940.
——. *Soil and Health.* Schocken Books, New York. 1947.
Sir Albert Howard developed the Indore composting system for tropical climates, and articulated the ecological viewpoint on the value and formation of humus.

Hyams, Ed. *Soil & Civilization.* Thames & Hudson, London. 1952.
"Those who do not learn from history are condemned to repeat it."

King, F.H. *Farmers of Forty Centuries.* Rodale Press, Emmaus, PA. 1911.
King was the head of Soil Management at USDA, and reports, somewhat romantically, on the careful attention paid to ecological cycles by the Chinese peasant.

Koepf, H.H., B.D. Petterson, & W. Schaumann. *Bio-Dynamic Agriculture.* Anthroposophic Press, Spring Valley, NY. 1976.
This is the most practical introduction to the Bio-Dynamic method available. The nature and history of the Bio-Dynamic movement are explained, showing how the metaphysical source of its theory is not at odds with solid farm practice.

Lampkin, Nicolas. *Organic Farming.* Farming Press. Ipswich, UK, 1990.
A comprehensive, British scientific textbook on organic practice, not theory.

Magdoff, Fred. *Building Soils for Better Crops: Organic Matter Management.* University of Nebraska Press, 1992
Clearly written, authoritative book giving research findings on the nature, role and function of soil organic matter, how it is built and how it is lost.

Martin, Deborah & G. Gershuny. *The Rodale Book of Composting.* Rodale Press, 1992.
A gardeners' guide to the principles and practices of composting.

Mollison, Bill. *Permaculture: A Designer's Manual.* 1990.
——. *Introduction to Permaculture,* with Reny Slay, 1991. Tagari Press, Australia.
The bible of sustainable agricultural design strategies.

Parnes, Robert. *Fertile Soil.* agAccess. Davis, CA, 1990.
An excellent guide to fertilization practice with somewhat radical emphasis on carbon.

Pfeiffer, E.E. *Soil Fertility: Renewal and Preservation.* Anthroposophic Press, Spring Valley, NY. 1943.

——. *Compost Manufacturer's Manual.* Anthroposophic Press, Spring Valley, NY, 1950.

——. *Chromatography Applied to Quality Testing.* Bio-Dynamic Literature, Wyoming, RI. 1959.

——. *Bio-Dynamic Farming and Gardening, vol. 1–3.* Mercury Press, Spring Valley, NY. 1983.

——. *Weeds and What They Tell.* Bio-Dynamic Literature, Wyoming, RI. 1981.

Dr. Ehrenfried Pfeiffer, a student of Rudolf Steiner, was a seminal influence on ecological agriculture in North America. His scientific rigor established a standard of excellence for ecological practitioners to follow.

Rateaver, Bargyla. *The Organic Method Primer.* Rateavers Press. San Diego, CA, 1993.

A new edition of one of the best books on organic gardening with a new section on plant absorption of nutrients that refutes ion exchange.

Reijntjes, Haverkort, & Waters-Bayer. *Farming for the Future.* Macmillan Press, London, 1992.

An excellent introduction to low external input and sustainable agriculture for the tropics with an invaluable annotated reference section.

Rodale Press Staff. *Controlling Weeds with Fewer Chemicals, Profitable Farming Now!* and *The Farmer's Fertilizer Handbook.* Rodale Press, Emmaus, PA.

These books are collections of articles from *The New Farm* magazine and special reports. They concentrate on a hands-on practical approach to regenerative agriculture.

Sarrantonio, Marianne. *Methodologies for Screening Soil-Improving Legumes.* Rodale Institute, 1991

A unique, practical guide to choosing the best legume for cover crop or green manure.

Savory, Allen. *Holistic Resource Management.* Island Press, Washington, DC, 1988.

A systems approach to managing resource systems. This method has been most widely implemented with intensive grazing.

Schmidt, H. & M. Haccius. *EEC Regulation "Organic Agriculture."* IFOAM Press, Tholey-Theley, Germany, 1993.

A comparative view of *Codex Alimentarius*, EEC, and USA regulations.

Schriefer, D. *From the Soil Up.* Wallace-Homestead, Des Moines, Iowa. 1984.

The best book available on the principles and practice of tillage.

Smith, Miranda, et al. *The Real Dirt.* Sustainable Ag Publications, Burlington, VT, 1994.
First-hand knowledge from more than 60 N.E. farmers on the many biological, cultural, mechanical, & even chemical tools available for reduced-input farming.

Steiner, R. *Agriculture.* Anthroposophic Press, Spring Valley, NY, 1928.
This is the record of eight lectures on agriculture given by Rudolf Steiner in 1924. He describes in metaphysical terms the interplay of terrestial and cosmic forces that create the processes of agriculture.

Storl, W.D. *Culture and Horticulture: A Philosophy of Gardening.* Bio-Dynamic Literature, Wyoming, RI, 1979.
One of the most readable and easy to comprehend explanations of Bio-Dynamic theory.

Sykes, F. *Humus and the Farmer.* Faber & Faber, London, 1949.
——. *Modern Humus Farming.* Faber & Faber, London, 1959.
These are practical books on biological soil management by an experienced and articulate British farmer.

Tompkins, Peter & C. Bird. *Secrets of the Soil.* Harper & Row, New York, 1989.
Strongly informed by Bio-Dynamics, an overview of the various fringe agricultural science practitioners, from Hieronymus to Perelandra.

Turner, N. *Fertility Pastures and Cover Crops.* Faber & Faber, London, 1955.
This British farmer concentrates on pastures, seed mixes, cover crops, and deep rooting herbs.

United States Department of Agriculture. *Soils and Men: The 1938 Yearbook of Agriculture.*
A wealth of information on soils, tillage, and fertilizer that should be reprinted. This is what the USDA was created to do before its shift to large scale agribusiness.

United States Department of Agriculture. *Report & Recommendations on Organic Farming.* 1980.
A landmark report, documenting the value and success of commercial organic farmers, as determined by a USDA study team. It called for further research into questions of interest to organic farmers, and has been ignored by the subsequent administrations.

Waksman, S. *Humus.* Wiley & Sons, New York. 1948.
——. *Soil Microbiology.* Wiley & Sons, New York. 1952.
The dean of American soil microbiologists. His earlier work on humus has not been surpassed.

Walters, C. & C.J. Fenzau. *An Acres USA Primer.* Acres USA, Raytown, MO. 1979.
A lesson-by-lesson approach to the techniques of ecological agriculture.

Wistinghausen, et al. *Bio-Dynamic Farming Practice.* 1990, Bio-Dynamic Agriculture Association, Clent, England.
A detailed, thorough text, useful to those seeking practical guidance in applying Bio-Dynamic principles on their farms.

Yeomans, P.A. *Water for Every Farm: Yeomans Keyline Plan.* 1954, 1993, Keyline Designs, P.O. Box 3289, Southport, Queensland, Australia 4215.
A dynamic approach to arid/dryland farming.

# Appendix D

## Texts & References

The following is a collection of useful technical reference books. Most are ecologically oriented. This is not an exhaustive listing, but a starting point for the serious researcher.


Alexander, M. *Introduction to Soil Microbiology.* Wiley & Co., New York, 1961.

Allen, M. *Ecology of Mycorrhizae.* Cambridge University Press, UK, 1991.

Ankerman, D. & R. Large. *Soil and Plant Analysis.* A & L Agricultural Laboratories, Inc. Undated.

Baker, K. & R. Cook. *Biological Control of Plant Pathogens.* Freeman Co., San Francisco, 1974.

Bartholomew and Clark, ed. *Soil Nitrogen.* Am. Soc. of Agronomy, Madison, WI, 1965.

Bear, F.E. *Soils in Relation to Crop Growth.* Reinhold, New York, 1965. also: *Chemistry of the Soil*, 1967.

Besson, J. M. & H. Vogtmann, ed. *Towards a Sustainable Agriculture.* Verlag Wirz A6, AArau. 1977.

Black, et al. *Methods of Soil Analysis.* American Society of Agronomy, Madison, WI, 1965.

Boeringa, R., ed. *Alternative Methods of Agriculture.* Elsevier Scientific Publishing, Amsterdam, 1980.

Brady, Nyle. *The Nature & Properties of Soils, 10th Edition,* Macmillan Publishing Co., NY, NY, 1990.

Chaney, D., et al. *Organic Soil Amendments & Fertilizers.* University of California, Division of Agriculture & Natural Resources, 1992.

Davies, B. *Applied Soil Trace Elements.* John Wiley & Sons, New York, 1980.

Dindal, Daniel (ed.). *Soil Biology Guide.* John Wiley & Sons, New York, 1990.

Durno, Moeliono & Prasertcharoensuk. *Sustainable Agriculture for the Lowlands.* Southeast Asia Sustainable Agriculture Network, Bangkok, Thailand, 1992.

Edens, et al. *Sustainable Agriculture and Integrated Farming Systems.* Michigan State University Press, East Lansing, MI, 1984.

Foster, R.C., A. Rovira, & T. Cock. *Ultrastructure of the Root-Soil Interface*. American Phytopathological Society, St. Paul, MN, 1983.

Golueke, C. G. *Composting: A Study of the Processes and its Principles.* Rodale Press, Emmaus, PA, 1972.

Hill, S. & P. Ott, ed. *Basic Techniques in Ecological Farming and the Maintenance of Soil Fertility.* Berkhauser Verlag, Basel, Switzerland, 1982.

Jackson, W.R. *Organic Soil Conditioning: Humic, Fulvic & Microbial Balance.* Jackson Research Center, 1993.

Khasawneh, F., et al., ed. *The Role of Phosphorus in Agriculture.* American Society of Agronomy, Madison, WI, 1980.

Killham, Ken. *Soil Ecology.* Cambridge University Press, U.K., 1994.

Kilmer, V., et al, ed. *The Role of Potassium in Agriculture.* American Society of Agronomy, Madison, WI, 1968.

Knorr, D. *Sustainable Food Systems.* AVI Publishing Co., Westport, CT, 1983.

Lockeretz, W., ed. *Environmentally Sound Agriculture.* Praeger, New York, 1983.

McAllister, J. *A Practical Guide to Novel Soil Amendments.* Rodale Press, Emmaus, PA, 1983.

McLeod, E. *Feed the Soil.* McLeod Publishing, Oak Park, MI, 1982.

Merkel, J. A. *Managing Livestock Wastes.* AVI Publishing Co., Westport, CT, 1981.

Miller, P.R., et al. *Covercrops for California Agriculture*. University of California Division of Agriculture & Natural Resources, 1989.

Mulongoy & Merckx, eds. *Soil Organic Matter Dynamics & Sustainability of Tropical Agriculture*. John Wiley & Sons, New York, 1993.

National Research Council. *Sustainable Agriculture & the Environment in the Humid Tropics*. National Academy Press, Washington, D.C., 1993.

— *Alternative Agriculture*. National Academy Press, Washington, D.C., 1989.

Pfleger & Linderman, Editors, *Mycorrhizae & Plant Health.* American Phytopathological Society, 1994.

Raven, P.H., R.F. Evert & S. E. Eichhorn. *Biology of Plants.* Worth Publishers, New York, 1992.

Russell, D. *Soil Conditions and Plant Growth.* John Wiley & Sons, New York, 1973.

Sanchez, Pedro. *Properties & Management of Soils in the Tropics.* John Wiley and Sons, New York, 1976.

Soil Science Society of America. *Nutrient Mobility in Soils: Accumulation & Losses,* [Special Publication No. 4]. SSSA, Madison, WI, 1970.

Stonehouse, B. *Biological Husbandry.* Butterworths, London, 1981.

Tisdale, Nelson, Beaton & Havlin. *Soil Fertility & Fertilizers,* (fifth edition). Macmillan Publishing Co, New York, 1993.

Van Wambeke, Armand. *Soils of the Tropics: Properties & Appraisal.* McGraw-Hill, Inc. New York, 1992.

Walsh, L. & J. Beaton. *Soil Testing and Plant Analysis.* Soil Science Society of America, Madison, WI, 1973.

Weaver, R. *Plant Growth Substances in Agriculture.* Freeman, San Francisco, CA, 1972.

# Index

**Bold page numbers**: the subject is discussed in detail.
*Italic page numbers*: the subject appears in a table or chart.

## A



## B



## C

## D



## E



## F



## G

# H



# I



# K



# L



# M

# N



# O

## U



## V



## W



## Y



## Z